Paradox Lost

Images of the Quantum

Springer
New York
Berlin
Heidelberg
Barcelona
Budapest
Hong Kong
London
Milan
Paris
Santa Clara
Singapore
Tokyo

Philip R. Wallace

Paradox Lost

Images of the Quantum

With 11 Illustrations

 Springer

Philip R. Wallace
Emeritus Professor of Physics
McGill University
104-1039 Linden Avenue
Victoria, British Columbia V8V 4H3
Canada

Library of Congress Cataloging-in-Publication Data
Wallace, Philip R. (Philip Russell), 1915–.
    Paradox lost : images of the quantum / Philip R. Wallace.
      p.   cm.
    Includes bibliographical references and index.

    ISBN-13:978-1-4612-8468-0   e-ISBN-13:978-1-4612-4014-3
    DOI:10.1007/978-1-4612-4014-3

    1. Quantum theory.   I. Title.
    QC174.12.W35   1996
    530.1'2—dc20   95-49921

Printed on acid-free paper.

Production managed by Hal Henglein; manufacturing supervised by Jeffrey Taub.
Typeset by Asco Trade Typesetting Ltd., Hong Kong.
Cover photo by Maria Pape/FPG International Corp.

9 8 7 6 5 4 3 2 1

SPIN 10526325

PROLOGUE

Some twenty years ago, I began offering a course at McGill University designed to introduce students of the humanities and social sciences to the problems and achievements of contemporary physics. Though this was by no means an original idea by that time, it met with some reserve and misgivings, since it was felt that if one said things in a way that could be understood, they would represent an inaccurate description of a complex subject, and that if one presented an accurate picture, it would not be understood. On the other hand, I, like many others of my generation, had my interest in physics aroused by popular writings of an earlier time, when Sir James Jeans and Sir Arthur Eddington tried to explain the then recent revolution in physics represented by the special and general theories of relativity, the consequent advances in cosmology represented by the "Big Bang" and the model of the expanding universe, and the sudden and rapid explosion of the quantum theory. These developments aroused some heated philosophical debates; I remember vividly the caveats expressed by the British philosopher Susan Stebbing in her excellent book *Philosophy and the Physicists*, whose sober analyses tempered some of the hyperbolic excesses and exaggerated extrapolations into areas outside of physics and even of science.

In more recent times, numerous "popular" books are again making their appearance, as are courses similar to my own, and again with a range far beyond the borders of sober science. Quantum mechanics in particular has engendered far-reaching philosophical speculations, some of which are highly anthropocentric, placing human consciousness at the center of consideration of the physical world. Librarians and bookstore owners have some difficulty deciding when these books should be shelved under "science" and when under "New Age" tracts.

I quickly discovered that a fair proportion of my students were already familiar with this literature and had formed from it some disturbing ideas of the essential character of the quantum theory. While I was trying to show them something of the extraordinary range of successes of the theory, they were confronting me with its alleged philosophical flaws, contradictions, paradoxes, and shaky foundations.

What was especially surprising was that relativity, which had created something of a philosophical revolution in the first two or three decades of this century, had gradually dissolved into the body of received and accepted knowledge. In the case of quantum theory, on the other hand, the arguments had hardly changed in sixty years, and the focus remained on *paradox* and a seemingly endless preoccupation with problems of *interpretation*. But in the intervening years, giant steps had been taken: Dirac had shown that the theory could be generalized to an operator form which embraced its special representations in terms of wave functions and matrices, and had also given it a manifestly relativistic form from which the idea of antiparticles sprung. Jordan and Wigner had pointed out the way to quantize all fields, from which came an understanding that the vacuum was not a void; the quantum "vacuum" was not empty but contained a residual energy of all existing fields. It had been shown that the properties of macroscopic forms of matter—and condensed matter in particular—were a result of quantum principles, as are the basic structures and phenomena of chemistry. Analytic methods for dealing statistically with macroscopic systems, through techniques of *density matrices* and *Green's functions*, dispensed with a physical role for wave functions and reduced them to formal analytic tools. New forms of matter with strange, exotic properties, like superconductors and liquid helium, were found. In the author's own domain of solid state physics, new coherent quantum states of matter proliferated—phonons, plasmons, magnons, polarons, helicons, and hybrids in which they interacted with each other to produce strange optical (and magneto-optical) forms of quantum excitation. Finally, at the level of the fundamental particles, as presently understood, altogether new physical properties, having no analog in classical physics, reflected the presence of strange symmetries (family relationships, if you will) at the root of the relationships of particles.

Is it not surprising that all this has not radically changed the character of the debate about the philosophical underpinnings of quantum theory and is scarcely reflected in the age-old debate about "interpretation"?

It was in this context that I found myself embroiled in the questions with which this book is concerned. It is said that age confers on people the right to become philosophers, though happily that right need not be exercised. My goal in this book is to make the nonscientific reader familiar with the present state of the subject and to put the "interpretation" debate in perspective. That debate, it must be understood, is not about the nature of the physical world; there are no conflicts between working physicists about quantum mechanics as generally understood and used. "A paradox is only a confusion in our own understanding," says Richard Feynman with his usual keen insight. The problem is not with the physical world, which is the concern of the scientists per se. The problem is in our minds—as difficult and uncertain a subject for the physicist as for the rest of humanity. It is the fact that the quantum theory can be formulated mathematically that makes it possible to keep this problem at arm's length for most of us; it is hard to argue with mathematics, while it forms a common base of experience from which to work.

When planning the agenda for my course, I chose to take as my starting point another revolution which occurred in physics in the mid-19th century. Classical physics ("mechanics") had over two centuries in which to be absorbed into human culture. The process was eased by the fact that it dealt with matters accessible to the human senses and was thus a part of the experience of everyone. But when Michael Faraday did his careful and exhaustive experiments, he found it necessary (as we all do) to create mental images of the invisible and intangible features of electricity and magnetism and of the relationship between them as a guide to an intuitive understanding of the subject. This led him to think not of mechanical objects but of "conditions of space"—of invisible "lines of force" carrying electrical and magnetic influences. We now call these "fields," and they were put into mathematical form by Maxwell, leading to his famous equations governing "electromagnetic fields." When it was discovered that these equations led to an understanding of *light*, their reality was beyond question, and a new perspective was introduced into physics.

There are similarities between the situation at the time of Maxwell and that of the early days of quantum mechanics. Both concerned aspects of the physical world that laid outside the sensory experience of man. Are electromagnetic fields physical entities or mere creations of our minds? For Maxwell, and the other physicists of his time, the electromagnetic *waves*, which provided a mathematical description of light, had to be waves of *something*, i.e., had to be material. Maxwell, using quite different mental images than Faraday, tried to give them substance in the form of the *ether*, whose properties (including invisibility) he tried for the rest of his life to understand. Maxwell was *interpreting* his electromagnetic waves, and it was not until Einstein introduced the theory of relativity in 1905 that his interpretation was proven to be untenable. In the intervening thirty years or so, before the interpretation could be discredited, enormous effort was expended to solve the problem of the properties of the ether. The phenomenon is strange. In retrospect, the ether was only a fabrication of the minds of physicists, yet there is no doubt that they felt that the problem was really a problem of physics.

It may become evident to the reader why the first two chapters of this book deal with the revolution of *fields* before embarking on the discussion of the quantum revolution. Ironically, it may now be argued (and is in Chapter 17) that it laid the groundwork for the new quantum mechanics, in that it opened up the possibility of considering that fields had superseded material particles as the basic elements of the physical world.

We shall then be concerned in this book with two themes. One is that of *interpretation* and the often paradoxical conclusions drawn from current interpretations; the other with the unfolding of quantum mechanics as a theory of extraordinary breadth and richness.

There is no lack of books purporting to tell the story of the quantum theory to the non-specialist reader. Most of them tend to emphasize the "strange" character of quantum phenomena; some discuss at length its "paradoxical" character. The problems discussed deal with the theory at its most basic conceptual level and exploit the reader's natural inclination to look for answers in their everyday experience of physical phenomena and the conventional interpretation given to it. Yet the evolution of quantum theory was driven explicitly by the discovery that these concepts were not adequate to explain phenomena presumed to have their origin at the atomic or subatomic level.

The sort of questions usually posed are, for example: What does the wave function really mean? What is the true nature of matter, and what is implied by "wave-particle duality"? Is reality created by the processes of observation or measurement? Does the wave function "collapse" when a measurement is made? Does quantum mechanics mandate instantaneous interaction between distant objects, thus violating the principle of relativity? Might there be "hidden variables" at work?

All of these questions depend on interpretations of the theory. Generally, the orthodox Copenhagen interpretation (itself often questionably interpreted!) is taken to be an integral part of the theory itself. A question seldom confronted is whether this is justified or whether the theory in its unambiguous mathematical form can stand alone, divorced from such interpretation.

Alternatively, the reader may be confronted with a plethora of possible interpretations and left to take one's choice without examining whether they are consonant with or in conflict with the mathematical theory (see, e.g., P.C.W. Davies' *The Ghost in the Atom*). So many different (and often conflicting) views are expressed that the reader is left with the impression that the theory rests in a sort of philosophical limbo, which can only mean that it is somehow philosophically flawed. However, philosophical disagreements, profound as they may be, do not stand in the way with respect to practical matters.

It is strange that the question of the meaning of an interpretation is usually avoided. Why do we not need to "interpret" electromagnetic waves, whose reality is attested by our radios and televisions, our microwave ovens and our medical X-rays? Can it be determined scientifically whether or not an interpretation is valid? What is the function of an interpretation? Does it provide insights into the full range of quantum phenomena or determine the methods employed to analyze them? Or is it simply a device for translating the precise mathematical structure of the theory into the language of classical physics, with all the ambiguities that that involves?

The reality is that there are *qualitative* and not merely technical or quantitative problems which transcend the philosophical issues arising from simple model problems. This is one of the major issues that will be addressed in the present book. The inadequacy, and even the irrelevance, of much of the usual philosophical discussion to current knowledge of *macroscopic* quantum systems soon becomes evident as these developments unfold; at the same time, newer and more sophisticated philosophical problems—problems more deeply embedded in the physics—take form. When the full scope of the theory in its most sophisticated form and the great variety of profound insights that it provides are revealed, the theory appears in a completely different light, not as a source of philosophical puzzles but as a revelation of how the whole of nature is ordered. It is a vast jigsaw puzzle in which every piece fits in an extraordinarily precise way. One cannot meddle with or change or "improve" it without confronting an awesome problem to which Feynman has repeatedly called attention: to find some other means to encompass all that we have learned with the theory as it is, as well as something beyond that which quantum theory does not adequately explain.

We will show that, at the level at which the theory is usually discussed, it does not give rise to "paradoxes" or philosophical difficulties that are inherent in the theory as used and understood by essentially all physicists. These difficulties reside rather in the imposition on the theory of interpretations that are not essential to its most precise form of expression. On the other hand, there is a wide range of quantum phenomena which reveal themselves in the macroscopic world of our senses and their technological extensions. We are, for example, familiar with lasers, superconductors, transistors, and the complex patterns of chemistry, often without being aware of their dependence on quantum principles, to say nothing of the manner in which quantum theory expresses itself in these phenomena.

The guiding principle behind this book is to be faithful to and be guided by the theory in its most precise (mathematical) form. It is at this level that the theory is generally accepted and used by working physicists. Mathematics is, as pointed out by Richard Feynman, a sort of shorthand language, but a language with its own intrinisic logic. The problem is to translate it into ordinary language without distorting its message. It is clearly no easy task to achieve the goals of this book without drawing on a background of ideas and insights not readily available to the untutored reader. On the other hand, to use a parallel with art, even if one cannot grasp all the nuances of the use of paint and brushes, or the tricks or illusions generated by the great artist in trying to draw us into his inner world, we can still stand back and look on in wonder, and in the depths of our soul feel a tingle of elation. Given the success and scope of the theory, unprecedented in the history of science, we propose a new perspective—a celebration of quantum mechanics.

CONTENTS

1

Introduction

In the third decade of the 19th century, Michael Faraday, son of a blacksmith with very little formal education, undertook extensive studies on electric and magnetic phenomena, which laid the foundation for modern physics. Great as his technical contributions to science were, his most important contribution was an intuitive idea—what we now call fields, *electric and magnetic conditions in space—on which contemporary physics is founded. From the time of Isaac Newton, physics had been the science of mechanics, the motion of material particles under the influence of forces. The idea of fields was not readily accommodated to such a framework, and a conflict of ideas ensued. Could fields be explained in terms of material entities? For this purpose, the idea of an all-pervading material ether was invented to carry the fields. Thus, a radical innovation in physics was accommodated by an interpretation, that of fields as material in character.*

Quantum mechanics provides a different vision of the world than does the familiar mechanics of Isaac Newton. The basic Newtonian principle is that things move, in a continuous fashion, according to the force exerted on them. If there were no force, they would stay still or move in a straight line with constant speed. The action of forces would make them accelerate (or decelerate) in a manner prescribed by his basic law:

Acceleration = Force/Mass

Acceleration is the rate of change in velocity.

We must distinguish between *speed* and *velocity*. The difference is not trivial. Speed is the distance traveled per unit of time (such as kilometers per hour) *regardless of direction*. Velocity is distance traveled per unit of time in a given direction. A car traveling at 100 km/hr in a northerly direction has the same speed, but a different velocity, from one traveling (say) east. The difference is important because, if force is exerted on a body that has, in its absence, been traveling in a straight line, its motion may change under the action of that force, either in speed or in direction; either is an acceleration.

Of course, reality is more complicated than that. Imagine a planet revolving around a sun. While it moves in its orbit, it may also be spinning on its axis. Thus, different parts of the planet are moving in different ways. Consequently, it is necessary, in dealing with macroscopic bodies, to apply the law of force to every part of the system; this, in turn, involves knowing something about the internal forces between its parts, as well as the external forces acting on it.

It is when we (as Newton before us) reduced the problem to that of the motions of individual parts of macroscopic systems that it became necessary to identify those component parts. The ancient Romans had already come to grips with that question but, having no observational or experimental evidence on the subject, they were reduced to speculation and a certain amount of metaphysics. They even recognized the possibility that the components of different forms of matter might have a different structure, but their models for the "atoms" of which matter was constituted were simply based on the concepts that governed their macroscopic view. The "elements" were merely macroscopic structures, reduced to the scale of the invisible.

We have here a first example of a problem that we face today with quantum mechanics: to deal with the elementary particles of physics in terms of the language and concepts of direct observation and experience at the macroscopic scale. All the evidence that we have concerning elementary processes is indirect. Our obsession with the conceptual difficulties and "paradoxes" of quantum mechanics has the same origin. This fact was emphasized by Bohr and led him to conclude that it was necessary to take account of the mechanisms of measurement to arrive at valid conclusions. Thus, hypotheses about measurement were incorporated into the interpretation of the theory though they did not appear in its structure.

Over a couple of centuries, the Newtonian program was successfully applied to many fields: astronomy, the dynamics of rigid bodies, fluid dynamics, the theory of sound and of musical instruments, elasticity, and so forth. After much travail, it was also applied by Maxwell to the theory of heat. It was only slowly recognized, however, that heat was merely molecular motion and that the problems could not be fruitfully dealt with by trying to follow the motion of all the atomic or molecular components of matter but rather by the use of statistical methods to develop a theory based solely on macroscopic properties —on the assumption that it was not possible to know *anything* of the motions of individual particles. These could then be assumed to be absolutely random within the constraints of general physical principles. How fundamental this approach is may be illustrated by the modern theory of *chaos* (*unpredictability*). More on this later.

Before introducing the idea of the quantum and its implications, there is another development that provides a part of the pool of concepts from which it sprung and on which it was built. This is the theory of electromagnetic fields evolved by Oersted, Michael Faraday, and James Clerk Maxwell. This arose from studies, with their origins in the distant past, of the "mysterious" properties of electricity and magnetism. Oersted, in the late 18th century, and Faraday, in the early 19th, discovered connections between electricity and magnetism that showed their close interdependence. Electric currents acted like magnets, and moving magnets produced electrical effects. Faraday introduced the concept of *electric fields* acting on electric charges and *magnetic fields* acting on magnets, including a compass needle. He also introduced the idea of *lines of force* along which the electric and magnetic fields acted and that permeated all of space. Thus, at any point, to use Faraday's own language, "A line of force may be defined as that line which is described by a very small magnetic needle, when it is so moved in other directions correspondent to its length, that the needle is constantly a tangent to the line of motion."[1] Translated into 20th-century language, what it says is this: If you put a fine-needle compass at any point, and move it constantly in the direction in which it points, the path it follows is the line of force through that point. Thus, lines of force pass through every point of space.

Faraday illustrated this by putting a magnet under a thin piece of cardboard and sprinkling fine iron filings over it. The filings became small magnets, which at every point showed the direction of the magnetic force (or *field*). In this way, he was able to visualize *fields*, though these fields were intangible and not material. This, as we shall see, turned out to be the central idea around which all modern physics now revolves. Elaborating on this idea, Maxwell, several decades later, was able to formulate a mathematical theory of electric and magnetic (and thus *electromagnetic*) fields and their interactions with matter.

Whatever hesitation there may have been to accept this radical new idea was quickly dissipated when Maxwell was able to use his mathematical equations to predict that, by virtue of the fact that varying electric fields could produce varying magnetic fields, which in turn produced varying electric fields, etc., electromagnetic waves could propagate themselves, by this sort of bootstrap mechanism, as waves in space. He was able to show as well that these waves would travel through space at the speed of light. What a marvelous moment this was for physics; not only were electricity and magnetism shown to be two sides of the same coin, but, taken together, they revealed the true character of light as an electromagnetic wave phenomenon!

Of course, 19th-century physicists were quite familiar with wave phenomena. An understanding of sound waves, water waves, and elastic waves in

---

[1] From *Lavoisier, Fourier, Faraday.* Britannica Great Books No. 45, p. 758 (1952).

solids all represented triumphs of Newtonian physics, which we shall hence-forth designate by the familiar term *classical physics*. Electromagnetism, how-ever, confronted them with a conundrum; all mechanical sorts of waves were characterized by oscillations in a medium (e.g., sound waves were oscillations of the pressure and density of the molecules of the air; water waves were distinguished by crests and troughs in the water surface, etc.). But in the case of electromagnetic waves, what was the carrier of the waves? What were they waves of?

The same sort of question was to arise again, as we shall see, in quantum mechanics. In the case of electromagnetism, the need to have a medium manifested itself in the invention of the "ether."

The idea of fields had no obvious place in the physics of Newton, though the major effort of physicists, including Maxwell himself, was devoted to the goal of giving it one. Faraday himself put it this way:

I desire to restrict the meaning of the term line of force so that it shall imply no more than the condition of force at any given place as to magnitude and direction; and not to include (at present) any idea of the nature of the physical cause of the phenomena; or to be tied up with, or in any way dependent on, such an idea. Still, there is no impropriety in endeavouring to conceive the method in which the physical forces are either excited, or exist, or are transmitted.

The thought was to explain (however tentatively) a new physical concept (fields) in terms of already existing concepts (particles and forces). He was, as the modern philosopher of quantum theory would put it, seeking an *interpreta-tion* of the new theory.

Faraday was astute enough, furthermore, to see that his new line of thought could incorporate, in retrospect, Newton's law of universal gravitation. One might ask, how could the gravitational pull of the sun on the planets reach out over the vast intervening spaces? By what mechanism could the gravitational force be transmitted? Once again, the explanation might be found in the *lines of force* reaching out through apparently empty space.

Subsequent "natural philosophers," as they were appropriately designated, were not so cautious or circumspect as Faraday. The idea quickly developed that Faraday's lines of force existed in a material medium, which was called the "ether;" this, it was postulated, was the medium that carried Max-well's electromagnetic waves. Such was the *interpretation* of electromagnetic fields.

But the ether seemed to defy direct observation; it could not be seen, or felt, or measured in any known way. Its properties could be inferred only from its context: It seemed to have no significant mechanical effect on bodies, like the celestial bodies, passing through it, and therefore must have very low density; it could sustain waves of incredibly high velocity, which, for a mechanical medium, would require enormous rigidity; it could transmit forces between bodies, such as those implied in Oersted's or Faraday's laws, in a direction perpendicular to the line joining them; and so forth.

To reconcile such seemingly contradictory requirements required great ingenuity. Maxwell, for example, understood that *transverse* forces would require torques or "twists" in the medium. But however great the difficulties, he never ceased to believe in the existence of the ether. He could not break himself loose from conventional interpretations.

The idea of fields did not bother him; he was already familiar with the use of a *field* viewpoint in the theory of fluids. The velocity of the particles at a certain point in the fluid was considered to be a function of position. Faraday's "lines of force" became "lines of [steady] flow." Maxwell made considerable use, in his electromagnetic theory, of analogies to fluid phenomena. He was, on the other hand, quite aware that this constituted argument by analogy and not by equivalence. Still, he could not rid his thought of mechanical images. Here is what he says in introducing his theory:

The theory I propose may therefore be called a theory of the electromagnetic field, because it has to do with the space in the neighbourhood of the electric or magnetic bodies, and it may be called a dynamical theory, because it assumes that *in that space there is matter in motion*, by which the observed phenomena are produced.[2]

The thrust of the work was thus to try to construct mechanical models for electric and magnetic phenomena. The "matter in motion" was the ether.

To the end of his days, Maxwell never renounced this view, which was commonly held in the scientific community. In an article on the ether that appeared in the 1898 *Encyclopedia Britannica* nearly two decades after his death, he states that

Whatever difficulties we may have in forming a consistent idea of the constitution of the ether, there can be no doubt that the interplanetary and interstellar spaces are not empty, but are occupied by a material substance or body, which is certainly the largest, and probably the most uniform body of which we have knowledge.

There may be a lesson here—that the makers of the great discoveries of science may not, in the long run, be the most reliable authorities on the true meaning of their discoveries.

A further note is worth adding. Faraday, the man without mathematical expertise, had another insight whose implications reached far into the future. Happily, in Faraday's time, it was not unacceptable to give voice to one's flashes of intuition. This led him to the following confession: "Final brooding

---

[2] From *The World of Physics* edited by J.H. Weaver, Vol. 1, p. 850. Simon and Schuster, 1987.

impression that particles are only centres of force; *that force or forces*[3] *constitute matter.*" Cannot one read into this simple sentence a premonition of Max Planck's photons, of Einstein's $E = mc^2$, and even of the meson theory of nuclear forces?

All of this shows that Maxwell's electromagnetic theory was born in a period of a shift in the conceptual framework of physics as profound as those implied in relativity theory and quantum mechanics. It reflects the same tension in the interplay of precise mathematical structures and the effort to reduce them to the conceptual framework that preceded them. The problem of "interpretation" is not peculiar to quantum mechanics. The historical witness should perhaps give us pause and lead us not to be overly dogmatic about interpretations, which, after all, define problems in our minds and not in the physical world. A more specific warning sign is to be found in the fact, acknowledged by Maxwell himself as a subject of concern, that the ether of his interpretation nowhere appears in his equations, as is the case for the "collapse of the wave function" in interpretations of quantum mechanics. The fruitful interplay of intuition and mathematically precise theories is better illustrated in the stories of Maxwell and Faraday than in most of the philosophical super-structure embodied in prevalent "interpretations" of quantum mechanics.

With this in mind, we shall approach quantum mechanics cautiously, at-tempting to stay within the bounds set by physical reality and the formalism of the theory and its implications.

------

[3] In the 19th century, the terms *force* and *energy* were often used interchangeably.

2

Beyond the Ether

Michelson and Morley proved that motion through the ether could not be detected experimentally. The validity of their conclusion was confirmed by Einstein's theory of relativity, which did not depend on the ether hypothesis. Thus, Einstein had proven that the "ether interpretation" was not tenable.

Efforts over several decades to construct a viable and credible theory of the ether led to consistent frustration and ended with the discovery that they were devoted to a misguided cause. In 1887, the American physicists Michelson and Morley carried out an experiment intended to detect the motion of Earth through the ether. The idea of the experiment was to detect variations in the velocity of light with the direction of its propagation through the ether. Maxwell himself had speculated on such an experiment, though he considered that, due to the very high speed of light, it would be an experiment technically very difficult to carry out. But the design and technique of Michelson and Morley's experiment permitted detection of the effect, if it existed, to very high accuracy. The result was negative; no such motion could be detected.

While continued efforts to sustain the ether concept continued, the special theory of relativity of Einstein (1905) showed a way out of the impasse. If there were indeed an ether, it represented, in Einstein's terminology, a preferred or absolute frame of reference, one in which the ether was at rest. The conception had arisen in the context of the Newtonian belief in the existence of an absolute space and an absolute time. Einstein's basic assumptions, that the laws of physics had the same form in all frames of reference moving with constant velocity relative to each other, and that consequently the velocity of light was the same in all such frames, left no place for the existence of an ether.

It also became evident that Maxwell's electromagnetic theory conformed to

Einstein's hypotheses; that is, its equations were invariant under transformations (called "Lorentz transformations") between the frames. The conclusion seemed evident—that Maxwell's waves, the waves of light—were indeed propagated through empty space!

The subsequent development of Einstein's general theory of relativity took the problem a step further. The special theory had proclaimed the equivalence of mass and energy; electromagnetic waves were waves of electric and magnetic energy. In the general theory, mass, and thus energy, caused warps or ripples in the fabric of space and time themselves. It seemed, therefore, that electromagnetic waves were not so much waves *in* space as waves *of* space!

At every step, it was necessary for physicists to follow the precept that Einstein put forward even for dealing with the moral problems of human societies, that it was necessary to change our fundamental ways of thinking. The old patterns of thought not only no longer worked; they often hid the realities of the world from us.

3

Introduction to Quantum Mechanics

Maxwell's theory of electromagnetic fields, which gave a mathematical form to the ideas of Faraday and others, demonstrated that light was a phenomenon of electromagnetic waves propagated through space. Such radiation is emitted by very hot bodies with a spectrum of frequencies, depending on the temperature. Max Planck showed that the frequency spectrum could be explained only on the assumption that the radiation was emitted and absorbed in elemental amounts called quanta, *whose energy was proportional to their frequency.*

Historically speaking, the idea of the quantum first appeared, in a very tentative form, in the work of Max Planck on the problem of the spectrum of black-body radiation. This is not an easy problem for nonscientists to grasp, so we shall not take this as the central line of approach. In broad terms, however, it concerns the fact that if a body is heated until it glows brightly, one can measure the intensities of the various parts of its frequency spectrum. Clearly, there is a finite amount of energy in the radiation. Insofar as the relative distribution of energies over the frequency range is concerned, at any given temperature, it rises from a very small value at very low frequencies to a peak which depends on the temperature and then decreases in an effectively exponential manner at high frequencies. The position of the peak in the frequency spectrum moves toward ever increasing energy as the temperature rises; in the visible spectrum, it rises from red heat to "white heat" before slipping into the ultraviolet and beyond. The major questions are, What determines where the maximum comes, and why do very high frequencies, as well as very low ones, appear so weakly?

Planck was able to show, empirically, that a solution could be found in the assumption that the light consisted of "bundles" or quanta of energy at each frequency, their energy being proportional to the frequency. Thus, the low-frequency end of the spectrum was characterized by quanta of low energy. At the high-frequency end, there was another problem: With increasing frequency, the quanta became so energetic that it would take an unduly high proportion of the total energy budget to excite them. Assuming, as turned out to be justified, that nature distributed energy equitably between its different modes and manifestations, the high-frequency, and thus high-energy, modes of electromagnetic energy demanded too large a share of the overall energy budget and thus were excluded.

In this way, Planck's hypothesis was able to explain the excitation spectra of the radiation.

Einstein quickly picked up the idea and applied it to other problems. One, for which Einstein won his Nobel Prize, had to do with the *photoelectric effect*, and it was here that one could see, in a most dramatic fashion, the difference between the classical view and the new quantum one.

We shall look at this phenomenon in detail.

4

ANALYSIS OF THE PHOTOELECTRIC EFFECT

Einstein undertook the analysis of the photoelectric effect in the light of the quantum hypothesis and showed that it explained accurately the results of experiments on this phenomenon, which were not consistent with the predictions of classical (nonquantum) electromagnetic theory. The reality of the quantum concept thus established, Niels Bohr undertook to use quantum principles to explain the existence and character of the discrete spectroscopic lines observed.

In its simplest terms, the photoelectric effect is the following: Ultraviolet light of a specific frequency is shone on the surface of a metal. Since light is an electromagnetic phenomenon, its oscillating electromagnetic field will act on the charged electrons in the metal. This process transfers energy to individual electrons. If these electrons are sufficiently highly excited in this way, they may have sufficient energy to escape across the metal surface, where their emission may be detected. It will then be possible to observe how the numbers and energies of the emitted electrons depend on the intensity and frequency of the incident wave.

Depending on the theoretical basis of the analysis, predictions can be made about the number and energy of the emitted electrons for a given intensity and frequency of the incident light beam. Predictions made on the basis of classical electromagnetic theory can be tested against experiment; as we shall show, these do not correspond to the results of experiment. Einstein, however, was able to show that if it were assumed that the electromagnetic field could transfer energy to the electrons only in discrete units (quanta) with energy proportional to the frequency of the light, the results observed could be explained.[4]

---

[4] In this context the nature or "interpretation" of the quanta does not appear to be at issue. This has been pointed out by D.C. Jones (Eur. J. Phys. **15**, 170–175,

We shall therefore look at the experimental facts and analyze them on the basis of both classical Maxwellian and quantum models and thereby test the validity of the quantum interpretation.

Fact 1: Below a certain frequency, no photoelectrons are emitted, however intense the beam of light may be.

- *Classical analysis*: The energy communicated to electrons will be proportional to the strength of the electric field; thus, if it is too weak, no electrons will be emitted. On the other hand, however low the frequency, if the beam is intense enough, the electric field should be strong enough to accelerate the electrons enough to escape. *The classical theory fails.*
- *Quantum analysis*: At low enough frequency, since the energy of a quantum is low, no electrons will gain enough energy to escape, no matter how intense the field is, i.e., no matter how many quanta there are. On the other hand, no matter how low the intensity, provided the frequency is high enough, the quanta absorbed will carry enough energy to allow the electrons to escape. The number of electrons emitted will, however, be greater the greater the intensity, that is, the greater the number of quanta. Therefore, *the quantum theory succeeds.*

Fact 2: The higher the frequency, the more energetic the emitted electrons are found to be.

- *Classical analysis*: When the frequency becomes too high, the emission of electrons will diminish. This is because of the inertia of the electrons. A high frequency means that the electric field, oscillating rapidly and thus alternating in direction very rapidly, pulls the electrons first this way, then that, so rapidly that they never have a chance to accelerate to the energies necessary for escape.[5] *The classical theory fails.*
- *Quantum analysis*: The energies of the quanta keep increasing with frequency and so then do the amounts of energy absorbed by the electrons that are emitted. *The quantum theory succeeds.*

Fact 3: There appears to be no delay before the emission of the electrons begins.

1994). But whether the quanta be particle or wave, the mechanism for the exchange of energy between field and electron remains to be specified. For the "particle" model, this is initially unspecified.

[5] A similar phenomenon can be observed if one fills a glass of water to within a centimeter or two from the top, and then stirs it back and forth with a spoon. At first, as you stir faster, the water surface will be more and more agitated and may spill over the edge of the glass. But if you stir *much* faster, it will be disturbed much less. The inertia of the water will not allow it to respond to the more rapid agitation.

- *Classical analysis*: The electric field of the radiation will gradually accelerate the electrons until they have enough energy to be emitted. Thus, there will be a delay in emission. *The classical theory fails.*
- *Quantum analysis*: The chance of a light quantum being absorbed in a given interval of time does not change with time.[6] Thus, the rate of emission will be uniform from the beginning. *The quantum theory succeeds.*

The conceptions of what is happening in the two cases are quite different, which puts into focus a very basic difference in the picture of physical processes in the classical and quantum theories. It is important to emphasize this because it warns us of the hazards of carrying over into quantum theory the mental images that classical physics provides. To be safe, it is necessary, in approaching the quantum theory, to purge our minds of previous patterns of thought, which, in the atomic realm, can lead us to completely false conclusions. Not to do so leads us to seeming "paradoxes," which are based on trying to use simultaneously incompatible images.

Let us summarize how quantum theory envisages the nature of light quanta (photons) and their interactions with charged material particles. First, let us note that light is a phenomenon of electromagnetic waves; this is made very clear by the fact that waves have characteristics, such as being able to interfere in space and time, that are common to electromagnetic waves and other waves (sound waves, water waves, elastic wave, etc.).

Next, we note that the energy of electromagnetic waves is quantized in discrete units proportional to their frequency; this amounts to a quantization of the *amplitudes* of the waves. This does not imply that these quanta are "particles" in the sense of pointlike objects; they remain waves, quantized or not, and thus are spread over space. They are, so to speak, the smallest possible components of electromagnetic waves.[7]

It might be considered merely a historical accident that the idea of the quantum first appeared in the context of electromagnetism and not that of the "material particles" basic to Newtonian physics. The fact that the spectral composition of the light emitted by various solids or gases was not continuous but showed evidence of sharp "lines" suggested that the energies of the atoms emitting the radiation were also quantized. This situation had already been identified as a puzzle, evidently not accounted for within the framework of

---

[6] This is, perhaps, a less than obvious conclusion from quantum theory. It has to do with the so-called "uncertainty principle," which we shall discuss in detail later (chaps. 10 and 12). It is not implausible, however, since the quanta are there from the beginning of the impact of the light beam on the metal.

[7] Bohr, in his Nobel Prize address in 1922, says that "In spite of its heuristic value ... the hypothesis of light-quanta, which is quite irreconcilable with so-called interference phenomena, is not able to throw light on the nature of radiation." Later developments showed that this statement is erroneous.

classical physics. Certain empirical laws governing the frequencies of these lines demonstrated the need for ideas to provide an explanation. But there was little hope of finding such clues until one understood better the nature and structure of the atoms themselves. The critical step in this direction was the discovery by Rutherford and his collaborators of the nuclear atom—the fact that the atom had a very small and very heavy nucleus, surrounded by a much more ethereal "cloud" of electrons thousands of times lighter. The image conjured up as an analogy was the solar system; just as the planets orbited around the sun, so the electrons, held by the inverse square law of electrical forces, would orbit around the nucleus. Once again, one had a classical base on which to lean, and Niels Bohr was the man to exploit it.

There were serious difficulties, however. One of the foremost was that it was known that when charged particles were accelerated, they emitted radiation (this was the mechanism for the emission of radio waves by an antenna). An electron in orbit about a nucleus was certainly accelerating, in response to the electric force holding it to the nucleus. A straightforward calculation showed that this radiation would be so strong that the electron would spiral into the nucleus in a minute fraction of a second; in other words, the "planetary" atom would be unstable. But this was no problem for Bohr; he simply suspended, by human decree, the laws of interaction between light and matter. The electron would not be permitted to radiate away its energy, and the orbit, of whose existence Bohr was confident, would be sustained!

There was a further problem: Orbits of any energy, and in fact of a continuous spectrum of energies, were possible. How, then, to determine *quantized* orbits? In a flash of genius, Bohr, noting that the famous Planck constant had the dimensions of an angular momentum, decreed (again) that the angular momenta of the allowed orbits must be integral multiples of this constant. In the simplest of atoms, that of hydrogen with only one orbiting electron, and if one allowed only *circular* orbits, he found it possible to obtain quantized orbits whose energies permitted an explanation of the main spectral lines of the light emitted when the atom had been excited.

This was a small accomplishment, which was treated with considerable satisfaction. The *Bohr atom* became one of the cornerstones of atomic physics. But problems remained. Sommerfeld showed how to extend the theory (still for hydrogen only) to elliptical orbits; this produced a splitting of the spectral lines. Very little real progress was made with the problem of multielectron atoms. The basic flaws of the planetary orbit model persisted.

5

DE BROGLIE AND ELECTRON WAVES

In 1924, Prince Louis de Broglie put forward the radical hypothesis that, in view of the fact that the electromagnetic radiation field manifested itself in quanta (particles?), electrons, previously thought to be particles, might have wavelike properties. This was subsequently verified by the diffraction experiments on electrons of Davisson and Germer. What, then, was the true character of electrons and photons (light quanta)? Could they be wavelike and still be "particles"? Max Born suggested a solution (an "interpretation"): They were really particles, and the waves were not themselves physical entities but served simply to determine the probability of the particle being at one point or another. The idea was eagerly accepted and has prevailed ever since. But Born himself still had some doubts.

This chapter closes with a discussion of waves and their properties, as groundwork for the idea of wave functions for matter.

In 1924, the young Prince Louis de Broglie presented a doctoral thesis in Paris in which he stated that "After long reflexion in solitude and meditation, I suddenly had the idea, during the year 1923, that the discovery [of light quanta] made by Einstein in 1905 should be generalized by extending it to all material particles and notably to electrons." Despite the greater authority of Bohr, Heisenberg, Schrödinger, and Born, de Broglie's idea was probably the most important, and certainly the most radical, single step made in the rise of quantum mechanics.

The linking of wave properties with those of classical particles by postulating relationships between classical dynamical variables (momentum and energy) and those of waves (wavelength and frequency) through Planck's constant provided a new principle applicable to all of the fundamental entities of physics. (See Section 5.1: Amplitude, Phase, and Interference.) In effect, it unified all physics by picturing the universe as made up of the interplay of many *fields* rather than of many point particles. This was to unlock the door to the ultimate expression of our modern understanding found in quantum field theory. It also maintained consistency with the tenets of relativity. Relativity implies relationships between energy and momentum different from those of classical physics. It requires that energy and momentum be related in essentially the same way as time and space coordinates. The latter linkage is a reflection of the fact that events that are interpreted as temporal in one frame of reference (like time dilation for moving objects in relativity) are interpreted as spatial in another (Lorentz contraction of moving objects). Similarly, "energy" in one frame of reference emerges as "momentum" in another moving relative to the first. Thus, quantizing energy requires quantizing momentum in a compatible way. At the same time, energy is quantized in terms of frequency, which expresses the rate of oscillation of a field in *time*, while momentum is quantized in terms of wave number (number of cycles of the wave in unit *length*).

Thus

$$\text{momentum in each direction} = \text{Planck's constant}$$
$$\times \text{ wave number in that direction,}$$

while

$$\text{energy} = \text{Planck's constant} \times \text{frequency of the wave.}$$

The novelty of such an idea, which seemed contrary to all intuition, was devastating to the whole philosophy of Newtonian mechanics. How was one to pose the problems of the physics of particles such as electrons if one could not follow them from point to point in time? There appeared to be no obvious way of translating that program into a world of waves. Electromagnetic waves were known to carry momentum and energy, but they were not localized at a point.

Once possible answer, which was soon rejected, was to be found in noting that Maxwell's equations seemed to provide a complete physical description of the physics of electricity and magnetism. The trouble with this was that the existence of the electromagnetic fields was evident only as providing forces acting on charged particles. It appeared that de Broglie's hypothesis implied a sort of dualism that removed the distinction between the forces and the objects of their actions. It was all too radical, and too confusing, to be digested and organized into a coherent conceptual structure. There was very little immediate evidence to support the new concepts or to give a clear view of their consequences. In the years to follow, widespread applications to a great variety of problems began to provide an intuitive base for them. At the beginning, however, it was as though one had been suddenly projected into an alien world where there were no familiar landmarks to guide one's progress; everything seemed paradoxical and puzzling.

It is hardly surprising, then, that physicists fell back on their old familiar picture of the world and tried to adapt the new reality to it. It had all started with the discovery that electromagnetism had manifested discrete units of energy. Did this not mean that there were particle-like entities in electromagnetism? This was in fact the line of thought that had guided de Broglie; if electromagnetic waves had "particle-like" properties, then why should not electrons and other "material" particles have wavelike properties?

It was scarcely acknowledged that the seeming symmetry was rather contrived. Photons of light were indeed discrete packages of energy, but they were not at all pointlike objects in space. But here there was a convenient dipping into the past. Newton had hypothesized a particulate character for light; it was embodied in the discipline of "geometrical optics," which dealt with light "rays" analogous to the trajectories of material particles. This discipline had been developed in a very sophisticated form by William Rowan Hamilton in the 19th century. It is true that this theory dealt only with light in special circumstances in which its wavelength was short compared with the dimensions of the objects with which the light interacted, so that the *wave* aspects were only weakly manifested; a small indication of a familiar landmark was welcomed. The fact that this was not evidence that light really had the character of point particles was conveniently overlooked. Rather, it reinforced the notion that matter was, after all, matter, at least in appropriate circumstances. If this were so, could one not once again picture it as having trajectories? It seemed that the gap separating the new physics from the old was being narrowed and that we could recapture the old pictures as a solid basis on which to construct a new order. It was in this intellectual atmosphere that attitudes toward quantum mechanics jelled, so to speak, and became relatively immune to basic disturbance from the ever-widening study of true quantum phenomena. The ambiguity continues to this day; *quantum field theory* and *particle physics* are interchangeable names for the same discipline! At the same time, so-called "wave–particle duality" has become an accepted part of quantum gospel, though it rests on an ambiguous use of words. We shall elaborate on this point later. The residue from this turn in the road was paradox, which still clouds the philosophical horizon today.

It was not long before de Broglie's hypothesis was put to an experimental test. The diffraction of light from crystals had been studied by the Braggs (father and son) as a tool for determining crystal structure and won them the Nobel Prize in physics in 1915. The idea was that the light scattered from the various individual atoms of the crystal created an interference effect. For a given wavelength of the light, the crystal could be oriented in such a way relative to the direction of propagation of the light beam that these scattered waves would interfere constructively. From this, the crystalline structure could be inferred.

Following de Broglie's suggestion, Davisson and Germer set out to see whether beams of electrons could produce the same effect. Not only would this provide a qualitative demonstration of the wavelike character of electrons, but it could verify the precise prediction of the postulated relation between the momentum, and hence the energy, of the electrons and the wavelength of their waves. The experiment confirmed precisely the de Broglie hypothesis.

It would seem obvious that the scientific community would have taken this as clear evidence that the field description of electrons was correct. But such was not the case. The idea of the electron as a *particle* was deep-rooted and not easy to shake. On the other hand, the evidence that it showed wavelike characteristics was unequivocal. The simplest reconciliation was to say that it was both; a viewpoint reinforced by the perception that this had already been shown to be the situation for photons. But how could one sustain such an ambiguous position?

Max Born, whose resolution of the quandary we shall discuss, has the following to say:

> Can we call something with which the concepts of position and momentum cannot be associated in the usual way, a thing, or a particle? And if not, what is the reality which our theory has been invented to describe? *The answer to this is no longer physics, but philosophy....*
>
> Here I will only say that I am emphatically in favour of the retention of the particle idea. Naturally, it is necessary to redefine what is meant.... The latest research on nuclei and elementary particles has led us, however, to limits beyond which this system of concepts itself does not appear to suffice. The lesson to be learned from what I have told of the origin of quantum mechanics is that probable refinements of mathematical methods will not suffice to produce a satisfactory theory, but that somewhere in our doctrine is hidden a concept, unjustified by experience, which we must eliminate to open up the road.[8]

That the answer is not to be found in modifying the mathematical structure is evident, since the theory as it has been developed, and as we will elaborate it, answers in a perfectly clear and unambiguous way the problems that we have submitted to it. There are no ambiguities, no paradoxes in quantum mechanics expressed in its mathematical form. Might it not be logical, then, to use this fact as a guide in our consideration of *interpretations*; that is, not to let "interpretations" in any way violate the theory in the form in which it is most consistent and clear? Is not the existence of the "paradoxes" that have occupied so much of our concerns simply a manifestation that we have, somehow or somewhere, violated that criterion?

As for Born, his resolution of the problem was to propose that the wave be considered not as a *physical* wave, as the electromagnetic wave was generally taken to be, but a "wave of probability." More precisely, the *intensity* of the wave—that is, the square of its amplitude at any point—was to be interpreted as determining the probability that the "particle" that the wave repre-

---

[8] Quoted in *The World of Physics*, vol. 2, p. 378.

sented would be found at that point. In this way, Born was able to retain the point particle within the framework of quantum theory. What was left unresolved, however, was a comparable interpretation of the other characteristic of the wave: its phase. The rather ambiguous term *probability amplitude* was used to characterize the wave function, but this again threw no light on the role of the phase. In this sense, the interpretation was incomplete. It also has the effect of differentiating quantum mechanics from all other physical theories, which have always been understood to provide a mathematical representation of real physical entities that could be translated into visual or conceptual "pictures" of elements or phenomena of the physical world. The difference is philosophically a profound one: that between a "reality" whose existence is the very raison d'être of the scientific enterprise and a symbolic representation of nature designed only to produce the results of experiments.

We shall see subsequently some of the puzzling consequences of the neglect of the phase of wave function, when we examine circumstances in which the phase plays a central role in the understanding of physical phenomena.

Born's proposition that refinements in mathematical methods will not suffice to produce a satisfactory theory seems reasonable. Beyond that, it does appear that we have let something in the door that is neither explicit nor implicit in the mathematical theory, something analogous to the ether in electromagnetism. What is it? Could it not be, quite simply, the concept of point "particles"?

Is that concept in fact "unjustified by experience?"

In the next chapter, we shall consider an example that will serve to illustrate the issues at stake. But first, we must clarify some basic terms.

5.1. AMPLITUDE, PHASE, AND INTERFERENCE

The major conceptual difficulty one faces in trying to understand quantum mechanics is that of replacing the idea, deeply rooted in our experience of physical events at the macroscopic level (where phenomena are directly accessible to our human senses), with the picture that best describes phenomena at the level of the elementary components of matter, on a scale a hundred million (or more) times smaller—the scale of atoms, molecules, nuclei, etc. Even though the Romans thought and wrote of *atoms*, conceived as the smallest components of matter, they did not conceive of the idea that this change of scale would involve a completely new set of concepts about what *matter* is. Their conception of the elements of matter was based simply on a reduction of scale; one could subdivide matter without changing its character, subdivide it again and again and again into smaller and smaller grains, until one reached a lower limit; *particles* could be smaller and smaller, but would still remain particles.

Thus, the concept of de Broglie that, at the ultimate level one found not localized specks but a diffused pattern of waves, was truly revolutionary. How, reductionists might ask, can what is clearly visible to the naked eye as a speck of material substance be found, on the atomic level, to be an ephemeral wave?

It was not that the idea of waves itself caused any difficulty. Over the centuries, wave phenomena of various sorts were familiar; but these phenomena were generally manifestations of large numbers of particles; they were, like the waves on a pond or a lake or a sea, a phenomenon of nature at the level of countless numbers of particles. That waves should again appear with the smallest possible components of matter seemed beyond belief.

Such a leap of the imagination did not occur, however, without the pressure of evidence. In the late 19th century, though scientific optimism was at its nadir, evidence was rapidly accumulating that our conventional conceptions were flawed and inadequate. Nearly a century later, we have been pushed further and further (and, one might say, deeper and deeper) into a view of the world as remote from our earlier concepts as the distant galaxies from our little planet. What has led us so far is the unimagined development of technologies. (Curiously, it was the influence of the technology of the industrial revolution that had given birth to the materialist view of the world that dominated the 19th century.) The modern technologies that have permitted us to look at the world at the level of what we conceive to be the ultimate components of our universe have compelled us once again to change our whole view of the world.

To find a wavelike character at the core of things is in fact a happy chance. The fact that we have some knowledge of, and experience with, waves is like finding familiar road signs on an alien planet; they give us a degree of confidence that we can understand our new surroundings.

We are familiar with many kinds of waves, from the waves we see on the surface of water to the sound waves by which we hear music and the electro-magnetic waves that bring us radio and television. Waves can have a compli-cated form, but in each case there are "pure" waves, oscillatory waves of a specific frequency,[9] from which all others can be produced. These play an important role in all wave phenomena; examples are waves of a pure tone in music or light waves of a pure color. However each sort of pure wave is produced, a periodic stimulus of a given frequency will produce a wave pat-tern, all parts of which will oscillate with the frequency imposed.[10] It is a

---

[9] By "frequency," we mean the number of cycles of oscillation experienced in a unit interval of time (usually 1 second).

[10] Strictly speaking, these remarks apply to systems that are *linear*; that is, systems in which effect is proportional to cause. The frequency of the generated wave is precisely that of its source. This is strictly true for electromagnetic or quantum waves; approximately so in most familiar circumstances for classical "mechanical" fields.

peculiar and surprising fact that when such waves of different frequencies are combined (i.e., superposed), however complicated the resulting patterns may be, these pure waves may pass through each other and emerge in their original form and with their original energy content. That is to say, each retains its identity.

Since water waves are easily visualized, they constitute a simple basis of discussion. Imagine that in a still pool of deep water we have a small float localized at some point, which we can jiggle up and down periodically, with any frequency we choose. The resulting wave will have an amplitude (which in a classical wave will determine its energy) as well as an oscillating phase, which will specify its instantaneous state in the oscillation. To adopt a different analogy, one might think of phases of the moon. The size of the illuminated moon goes through cycles, from full brightness to near extinction and back again. Where it is at a certain point in this process is called its *phase* at that point. One can make an analogy with going around in a circle, where one returns to the initial point after a certain time or after having gone a certain distance. Since in the circle one goes through an angle of 2π radians, a change of phase of this amount is equivalent to a cycle. A change of phase of 2π is equivalent to no change at all. A change of phase of π corresponds to the change from trough to crest or the inverse.

As for the amplitude of a water wave, it describes the magnitude of the periodic oscillations. If the stimulating source is stronger—that is, if it is jiggled twice as energetically—a wave will be created with exactly the same pattern but with twice as much energy. It will simply be, so to speak, magnified. The energy in the wave is proportional to the square of the amplitude of the vibrations; in this case it will be increased by about 41%. The important feature here is that, in "classical" waves, without changing the frequency, we can increase the energy and amplitude of the wave continuously by increasing the energy input. In quantum mechanics, however, since Planck's law specifies that the energy of a quantum is proportional to the frequency, the wave amplitude must be frequency dependent.

What has this to do with the "probability interpretation"? In this interpretation, the probability depends only on the *amplitude* of the wave function; no mention is made of an interpretation of phase. Yet phase correlations between states are essential to the description of quantum transitions and thus of all physical change. Furthermore, the probabilities relate to a supposed "point particle," and fields are supposed to tell us something, not about the behavior or properties of this particle, which nowhere appear in the theory, but only about its location. Pagels, for example, in *the Cosmic Code*, says that the electromagnetic field, with its energy quantified, gives only a probability distribution for the photonic "particle." It gives no physical description of a time evolution of the particle, and the electromagnetic field is denied objective physical reality. As for what the particle "does," or why, the only answer we are given (first by Feynman) is that "nobody knows." I have repeatedly met the same response in defenders of the conventional interpretation. It is in this sense that I believe that it must be said that the interpretation (but not, as Einstein would have it, the theory) is incomplete.

Suppose now that we complicate things a bit. We create a barrier to the passage of the wave; some sort of breakwater. The waves will then reflect back from it. But suppose that we make two small holes in the breakwater, with a distance between them neither much larger nor much smaller than the wavelength of the wave. Then we would find waves emanating from each of the holes. Shortly, we would have these waves crisscrossing each other, making a more complicated pattern of crests and troughs. Closer study would reveal that this pattern would be made by simply adding, at each point, the amplitudes of the two separate waves at that point. This is the phenomenon of *interference*. Now, of course, the frequency of this more complicated pattern would still be that of the original waves. What we will have created is a more complicated wave of the same frequency. If we were to place a row of lighted buoys at some distance behind the breakwater, we would detect an *interference pattern*. We can see what the general pattern would be. If we were to pick a point on this "detector" at an equal distance from each hole in the breakwater, we would find a strong peak in the wave there. If we move away from that point such a distance that the difference of its distances from the holes is half a wavelength, the vertical displacements of the two waves will add up to zero. If it is a whole wavelength, we again get an amplitude peak. This pattern will repeat itself for each displacement that changes the difference of the path lengths from the opening to the point in question. Note, however, that *these* peaks and valleys represent limits of the *amplitude* of the combined wave; they represent the extreme limits of the periodic oscillation. Another way of putting this is to say that there is still *phase coherence* in the wave; the pattern of displacements is changed, but at all points the vibration takes place periodically, with everywhere the same period.

To recapitulate: the *intensity* of the wave, which is a measure of the amount of energy it carries, is determined only by its amplitude. Changes of amplitude retain the pattern of the wave but change its scale. This is true for both the classical and quantum waves, regardless of the frequency. There is still a definite temporal phase proportional to the frequency, however, for the overall pattern of the wave. This is true of all states, in whatever circumstances, provided that they have a definite frequency. If, however, we add waves of *different* frequencies, this will no longer be so. In this case, each of the component waves will carry its own energy, regardless of the overall pattern. In this sense, each retains its own identity. The overall pattern itself is not stable but changes with time. On the other hand, by combining waves of the same frequency, we can create stable patterns with that frequency. This will include waves that go through two separate holes. In that sense, the parts going through the two holes do not have a separate identity but constitute a single wave.

One must be careful to be clear about the meaning we give to the word *interference*. We shall take it to mean the process of combining waves of the same frequency to form others of the same frequency. For clarity, when two or more waves of different frequencies are combined, we shall characterize it not as an interference pattern but as a superposition pattern.[11]

---

[11]This is important to the assertion of Dirac (*Quantum Mechanics*, page 9) that "Photons (light quanta) cannot interfere with each other, but only with themselves." It will also be important in understanding the stellar interferometer of Hanbury Brown and Twiss (see chapter 20).

6

The Wave Function and Feynman's Two-Slit Experiment

Wave–particle duality, still being debated 70 years later, was put into focus by a "thought experiment" proposed by Richard Feynman. If a beam of electrons strikes an obstacle with two narrow parallel slits through which they can pass, the interference between the parts of the beam passing through them is known to create an interference pattern on a detecting screen behind the slits. But there is a paradox here: Since individual quanta are absorbed by the detector, to show interference each would have to pass through both slits. But how could they do that if they were particles?

We suggest a different way to resolve the paradox: to acknowledge that the "particles" are actually quanta of waves and so are not localized. This interpretation is entirely consistent with their mathematical definition.

Richard Feynman, in his excellent Messenger Lectures at Cornell University in 1964, published under the title "The Character of Physical Law," devised a simple experimental setup that seemed to him to illustrate typical quantum behavior. First, there is a source for a beam of either electrons or photons. This beam then impinges on a screen in which are pierced two narrow parallel slits. Beyond the screen, and at a distance from it that is large compared with the distance between the slits, is placed a detector, which could be a sensitive screen capable of recording the absorption of single quanta of energy. The beam is weakened sufficiently that individual events are detected. Feynman says, perceptively, that any other situation in quantum mechanics can be explained by saying, "You remember the case of the experiment with the two holes? It's the same thing." The experiment demonstrates "typical quantum behavior," so that we do not need to depend too much on the precise details but can merely concern ourselves with principles that have been very widely verified.

Whether we use beams of electrons or of photons, the behavior is essentially the same. At first, the individual absorption events will appear to take place randomly; a dot here, a dot there, indicating the absorption of a single particle. But as the events accumulate, the vague outlines of a pattern will appear. With more and more events, the pattern will become increasingly evident. It is a bit like a newspaper picture, which is made up of tiny dots. The picture will reveal itself with the accumulation of events.

The pattern that appears shows a typical interference effect. There will be alternate regions of dark and light, fading continuously into each other. Similar effects can be seen with other kinds of waves. We have illustrated this with water waves in Figure 1b.

It is worthwhile to reflect on what is going on. You cannot, at any point, separate the two displacements whose sum the pattern reveals. Individual molecules of water do not "belong" to one wave or the other. But something extraordinary may be observed, as if by magic. If the two waves are of finite length, they will pass through each other and emerge on the other side with their separate identities intact! This is most easily observed when the waves are one-dimensional, like those on a stretched string or a slinky toy.

Is there a true parallel between the classical and the quantum cases? With water waves, the interference manifests itself as a continuous phenomenon. In the quantum case, it appears only statistically, as the accumulated effect of a number of seemingly independent events. These events correspond to the absorption of independent electrons or light quanta (photons), as the case may be. But if they are independent, how can one explain the fact that, in the end, they combine to form a pattern? Where is the construction plan for the pattern to be found? One is tempted to imagine a parallel with the relation between human genes and the organs or functions for which the genes somehow carry the programs. The implication seems to be that each one of them carries all the information needed to make the pattern! Why then do they register only at specific points?

The answer to this involves two elements. First, we must be careful not to confuse the spot on the detecting screen, large enough to be detected with the naked eye (or some other macroscopic instrument), with a geometric point. On the atomic scale, the registered "dot" may have the dimensions of many atoms. If it were really a point particle that was detected, its impact or absorption would have to have been dramatically amplified, even if it appears like a point. The second factor to consider is that what has taken place in the process of absorption is an elementary atomic process, which will take place locally *on an atomic scale*. Similarly, if light quanta (photons) are used in the experiment, they will most likely be absorbed by one electron in the detector. That one electron may then give up some of its energy to other particles, but that is a secondary process. But the "absorption at a point" does not mean that the electron or photon itself is a point object, only that what it has

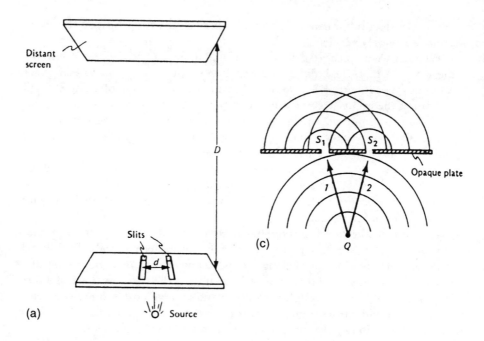

Distant screen

D

Slits

d

(a)   Source

S_1 S_2

Opaque plate

1 2

(c) Q

interacted with on the screen is an electron localized on an atom or molecule of the detector. The mathematics of the problem describes the absorption of a *wave*.

In the case of photons, it is important to note that a photon is by definition a quantum of the electromagnetic field. In classical electromagnetic theory, a wave goes through the slits and is diffracted, giving rise to interference. The quantum is a *sample* of the wave with an amount of energy proportional to its frequency. Frequency implies extension in time; the connection of time and space required by the theory of relativity then makes extension in space necessary. This in turn requires that the single photon, being a minimal sample of the wave, is also subject to interference. The reason that it cannot reflect the interference is that its totality must be absorbed locally. It interacts with all electrons, but only one of them can enter into a quantum transition with it. What determines which one? There are an enormous number of possibilities; the matter is one of pure chance. However, the probability of interacting with any one will be proportional to the intensity of the wave at that electron.

The situation is similar for an electron beam. Field theory, as we shall explain later, enables us to think of an electron as a quantum of a physical electron wave. In this sense, the situations with respect to electron and photon are symmetrical. But as we shall see in the next chapter, there can be no classical electron wave, since only one electron can be in each possible quantum state.

To summarize: Each particle in the beam, whether a light quantum or an electron, is an indivisible thing that goes through both holes and interferes only with itself. It can interact only as a whole.

Feynman, however, posed the question, Can we not test whether an electron goes through a particular hole, by placing behind that hole a source of very intense light? If an electron goes through that hole, you will be able to "see" it because the light will be scattered off it. You can then see if the interference pattern is changed because of that. What is found is that something is "seen" behind the hole in question, and the interference pattern is weakened by it. The moral seems to be that if an electron is found to have gone through that hole, it does not contribute to the pattern. Does this not prove that the electron can go through a single hole?

This seems again to introduce a paradox: There can be interference only if the electron goes through both holes; yet, if one does not catch the electron going through one, it goes through both, as shown by the interference pattern.

◄───

Figure 1.
(a) The proposed experimental setup: source, screen with parallel slits, and detecting screen. (b) An analogous water wave experiment, with wave interference. The line shows how the interference pattern would look, with alternating light and dark bands. (c) A schematic representation of how spherical wave patterns would spread out from the two slits. (From D. Falk, D. Brill, and D. Stork, *Seeing the Light*. Harper and Row, New York, 1986.)

From the quantum mechanical viewpoint, the matter can be put differently. One way is to use the type of argument proposed by Heisenberg, who says that the electron can be detected only "at a point" (or in a given small neighborhood) by using a photon of very small wavelength and thus very large momentum or energy. This completely throws the electron off track, giving it a large momentum change and thus short wavelength, which destroys the interference. But of course it is not just a *part* of the photon that is scattered; it can only be scattered as a whole. Does this not prove that the *whole* went through the given slit? But if the whole photon went through it, it would not show an interference effect anyway. So we are left with the curious proposition that it went through the one slit *because* we tried to measure it doing so. But how would it "know" that we intended to measure it?

This is precisely the sort of argument that is often put forward to show that quantum mechanics depends on human intervention, etc. But such a course seems to represent an abandonment of the whole scientific viewpoint and so has very disturbing consequences, which could only be accepted *in extremis*. Fortunately, a simple argument is possible that rescues us from the dilemma by changing the ontological landscape. Let us put aside any argument that is in any way dependent on the notion that the electron is or acts like a point particle, and let us accept de Broglie's hypothesis at face value—that the electron, like the photon, *is* a wave, in a state described by its wave function. Then it must, unequivocally, go through both slits. What, then, of its detection by a photon? We now simply describe this as a quantum transition from one state to another. This means that the whole electron wave function is transformed into another state with a different wave function. The source of this new state is situated behind the hole where we have placed the detecting beam. Because of the high photon momentum, the electron will be given an enhanced momentum, so it will no longer have the character that enables it to show an interference effect.

There is nothing peculiar or abnormal about such a transition. Can an extended wave bring about a *local* transition? Certainly, yes; that is what happens when an electromagnetic field brings about a transition of an electron bound in an atom from one state to a quite different one or even frees the electron from the atom altogether (the process of ionization).

The preceding discussion differs substantially from one based on Born's probability interpretation. Where probability enters into our discussion, it relates to the randomness of the electron that interacts with the incoming electron or photon; this in no way associates probability with the wave function itself, but rather with its interactions. What, on the other hand, are the implications of the probability interpretation? According to that interpretation, probabilities could be associated with whether the particle would go through one slit or the other. But if it went independently through one or the other, no interference pattern would ensue. Interference implies a correlation between the two probabilities, a correlation then at a distance. And that correlation would affect the behavior of each individual particle observed.

The reality is that probability concerns only the amplitude of the wave function, whereas the interference is strictly a phenomenon of the phases. Interference is certainly a physical phenomenon; it is hard to conceive of interference of probabilities, since probability is only an intellectual concept, an epistemological tool for the analysis of phenomena. Traditionally, we think of theories as statements of the relationships of physical entities and phenomena, not of our methods of analysis.

It is through this confusion that the human observer and the process of measurement have insinuated themselves into physical theory. There is not an iota of evidence that humans have any special role in physical phenomena; there is no credible evidence that our interactions with the physical world are not subject to physical law. Further, there is no "human factor" to be found in the mathematical structure of quantum mechanics, just as there was no "ether factor" in electromagnetic theory.

7

THE PAULI EXCLUSION PRINCIPLE:
THE IDENTITY OF PARTICLES

Two subsidiary principles governing quantum particles were found necessary to make quantum theory work: (1) the Pauli Exclusion principle and (2) the principle of identity of all particles of a given type. For purposes of the Pauli principle, particles are divided into two categories: fermions, characterized by half-integral spins, and bosons, characterized by integral spin (both in terms of ħ). In the case of fermions, no two particles can occupy the same quantum state, but this is not the case for bosons, for which occupancy is unlimited. As for identity, this applies to all sorts of particles. Identity is a tricky concept with some startling consequences; identity does not mean simply that they are copies of one another, like identical twins, but that they are indistinguishable in the sense that two dollars in a bank account have no separate identities.

Of course, there are very important differences between electron waves and photon waves. Photon waves do not carry any electric charge, although they interact strongly with electric charges. Electron waves do carry charge and thus are acted upon by the charge of the nucleus. There are other differences as well. Electron waves have the characteristic, embodied in the *Pauli exclusion principle*, that no two of them may exist in precisely the same state; the same is not true for photons. Many identical photons can make up a macroscopic electromagnetic field. Because of the exclusion principle, there can be no *macroscopic* (i.e., classical) electron field. But this point does not touch on fieldlike behavior per se, only on the distinction between two sorts of field.

If there are differences, there are also similarities, one being in the fact that all photons are indistinguishable from each other; the same is true of electrons. This is a much more profound and subtle point than one might at first imagine, and its consequences are important. It is not the same indistin-guishability as that of two identical twins, because we can still mark them so as to distinguish one from the other. But for quantum particles, this is not possible. It is the indistinguishability between 4 obtained by adding 1 and 3 from one obtained by adding 2 and 2. It is the indistinguishability of two dollars in one's bank account.

How can one know whether or not they are distinguishable? Let us ask in how many ways we can take $4 and divide it between two bank accounts. Let us list all the possibilities:

	Account 1	Account 2
A	4	0
B	3	1
C	2	2
D	1	3
E	0	4

So there are five ways it can be done.

Now consider four balls of different colors: red, green, blue, white. In how many ways can we apportion them between two boxes? While there is still only one way to realize cases A and E, there are 4 ways of realizing B and D and 6 ways of realizing C, so there are now 16 ways rather than 5 of accom-plishing the task.

This can be important in thermodynamics, where the entropy of a system depends on the number of ways of realizing it. Thus, the thermodynamics of identical particles is quite different from that of distinguishable ones.

This distinction will play an important role in certain well-known para-doxes of quantum mechanics, where it is the unjustified implicit assumption of distinguishability that is the origin of the paradox.

8

The Schrödinger Wave Equation and the Hydrogen Atom

In 1925, Erwin Schrödinger, drawing heavily on the experience of classical waves, introduced a wave equation for electrons, from which one could deduce their states and properties (energy, momentum, angular momentum, etc.). He applied this to the hydrogen atom with great success, determining the energy and angular momentum states of the hydrogen electron. From the energy states, one could determine the spectral frequencies; the decay of an excited state induced by heating to the ground state was accompanied by the emission of radiation of quanta whose energy was equal to that lost by the electron, and from this, the radiation frequency followed from Planck's law. He thus correctly predicted the major spectral lines of hydrogen.

Several residual problems of the Bohr theory of the atom were thus resolved. Radiation collapse was avoided by the discovery of a state of lowest possible energy, and states of zero angular momentum, inconceivable in Bohr's scheme, were found.

De Broglie himself only dealt with "free" electrons, which he described in terms of simple plane waves. The problem of atomic structure, on the other hand, required the study of how such waves would behave in the presence of not only the nuclear force on the electrons but also the repulsive force between the electrons themselves. De Broglie was aware of this problem, but it was Erwin Schrödinger who found its resolution. In a sense, the equations governing the behavior of other kinds of waves served as a model for the postulation of a *wave equation*. The question was how to incorporate the Coulomb attraction of the electron to the nucleus into the equation. For all wave phenomena where the waves are subject to constraints, it is found that solutions to the wave equation that satisfy these constraints exist only for special values of some parameter (usually the frequency or wavelength). Such is the case, for example, for the waves on a vibrating violin string, which is fixed at its two ends. The wavelengths of the allowed modes must be such that there are nodes at the two ends, which means that the length must correspond to an integral number of half-wavelengths.

Similar problems exist for vibrating drumheads and sound waves in organ pipes, electromagnetic waves in waveguides, etc. In all cases, there are allowed modes of vibration with definite frequencies (frequency being a function of wavelength). Because of Planck's hypothesis for quantum systems that energy is proportional to frequency, the object of Schrödinger's equation is to find the energy states of systems, in this case, the hydrogen atom. The constraint in this case is not a boundary but the attractive force of the nucleus, which binds the electron to it.

Since the problem is that of determining the allowed energy states, it is the potential energy of the Coulomb attraction that must enter the equation as a contribution to the energy.

When the energy spectrum of the electron in hydrogen is calculated using Schrödinger's equation, it gives values that are, aside from minor auxiliary effects, in correspondence with spectroscopic data. But aside from giving the energies, the equations yield the patterns of the wave functions that correspond to each of these energies.

In mathematical terms, the overall structure of the problem for electrons is similar to that for electromagnetic waves. In the latter case, we do not (or did not until recently) ask, What does it mean? What is this pattern a pattern of? We do not hesitate to say that it is a pattern of the electric or magnetic field; in the electron case, on the other hand, it is not generally considered adequate to answer that it is a pattern of the electron field. Only chemists, who tend to be pragmatic types without philosophical pretensions, seem to regard this answer as sufficient. Chemists of the present day generally do not think of point electrons spiraling about the nucleus but think of *electron clouds*, which are fieldlike entities.

How does the picture of the hydrogen atom based on Schrödinger wave functions differ from that implied in Bohr's earlier "planetary orbit" model? The first question that comes to mind is, Why is the hydrogen atom stable in this model, whereas it was unstable in Bohr's? The simplest answer is that in the quantum mechanics of Schrödinger's atom the electron is not *accelerating*. This directly reflects the fact that it is not following a trajectory as a classical point particle does but has a wave character. The clouds of electrons are not circulating around the nucleus. In the lowest ("ground") state, the clouds do not even have any angular momentum about the nucleus, unlike in the orbit in Bohr's picture. In most of the excited states, the electron does in fact have an angular momentum, though it is in a steady state. How can something static carry an angular momentum? The answer could only be in the wavelike character of the electron, characterized by amplitude and phase. The situation is similar to that of the charges in a loop of wire in which flows a steady current. Though there is constant current flow, nothing changes with time. Both charge and current distributions are constant and independent of phase and give rise to constant electric and magnetic fields. The magnetic field is generated by the current and thus is associated with the angular momentum. The existence of (quantized) angular momentum about the symmetry axis, which depends on the phase, therefore does not affect the electron density, which is determined by the amplitude of the wave function.

If the hydrogen electron is in an excited state, however, it may interact with the electromagnetic field to radiate a photon, that is, a quantum of an oscillating electromagnetic field, while dropping into a lower energy state. The wave for the initial state has a different energy and thus different frequency than that of the final state; the wave function during the decay will be in a "mixed state," which is partly initial state and partly final state. The interference between these two creates an oscillating phase, which stimulates a correspondingly oscillating electromagnetic field, whose quantum is one photon of an energy that is the difference of the initial and final state energies, so that conservation of energy is maintained.

We see here the quantum analog of the spiraling in of Bohr's orbiting electron. Suppose that we have a hydrogen atom in a highly excited state; it may then radiate a series of photons, one for each drop in the atomic energy levels through which it passes. Rather than radiating a continuous spectrum of radiation, it radiates a series of photons of ever-increasing energy until it settles stably into its lowest energy state.

What is it that sustains this lowest energy state? It is the wavelike or fieldlike character of the electron, which implies that it is a nonlocal entity. A simple way of looking at it is in terms of the so-called *uncertainty principle* of Heisenberg. As the electron gets closer and closer to the nucleus, it is more and more localized in space. But this means shorter wavelength (or, more generally, more rapid spatial variation). Shorter wavelengths imply higher frequencies and, thus, higher energies. These ultimately compensate the energy loss that is due to proximity to the nucleus, and the electron ends up in a stable state where these two tendencies are in balance.

9

CRITIQUE OF THE POINTLIKE ELECTRON

Born's probability interpretation did not fully resolve the problem of the meaning of the wave function. We have noted that waves are characterized by two quantities: amplitude and phase. For wave functions, the amplitude is interpreted as determining the probability distribution of a point particle. The phase, on the other hand, is responsible for interference effects. But just as Maxwell's ether did not appear in his equations, the point particle does not appear explicitly in the wave equation. No physical mechanism is proposed to explain the probability distribution assumed. Nor is it clear how to attach a phase, a natural characteristic of waves, to a point particle.

A different light is thrown on the problem by the approach of Feynman, which will be discussed in Chapter 16.

Max Born's feeling that keeping the image of the point particle was essential led him to propose an interpretation that, it seemed to him, dealt with the inadequacies of the wave picture. He proposed that the wave function not be considered as a physical reality but only as an indication of our knowledge about the electron. In particular, he proposed that the wave function merely determined the probability that the electron—a pointlike object—was to be found in one position or another. This probability was determined by the square of the *amplitude* of the wave function.

Strangely, the fact that this interpretation ignored, or found no place for, the phase did not appear to bother the physics community, which readily accepted Born's idea. This was no doubt due to the fact that it seemed at that time that the phase was of very little physical interest; it was only much later that macroscopic quantum phenomena based on the phase coherence of many-body systems were brought to light. The same is true of Feynman's *path integral* viewpoint (more on this later), which was yet to come into prominence, with its emphasis on the interference of phases of different paths. In any case, by the time these developments had appeared, statistical methods for many-body problems had come into prominence, and these dispensed with the need for consideration of wave functions, and the question of interpretations had gone into eclipse in the mainstream of physics. Interestingly, however, these new methods used field-theoretical techniques, and formalism had largely displaced image in the milieu of current physics.

Other phenomena seemed to most physicists like manifestations of a point electron but mostly because this image was deeply rooted in advance. For instance, in the famous *gedankenexperiment* of Richard Feynman, a beam of electrons passing through two slits were reunited to give an interference pattern on a detector. But when the intensity of the beam was reduced, it was possible to see individual dots on the screen, which were interpreted as manifestation of localized electrons. When enough of the dots were collected, they did in fact show the interference pattern typical of interfering waves. Since the individual dots seemed to occur randomly, it was assumed that the location of the electrons was random and reflected the probability distribution of the location of electrons. But there was an alternative and perhaps more realistic interpretation. Since photons had to be absorbed whole, and this would normally be done by one particle or one atom that was itself localized within atomic dimensions, the localization might be a characteristic of the detector rather than of the detected electron. Would it then be legitimate to think of the absorption of the electron as the gradual absorption of a macroscopic wave? A calculation of the process does seem to describe exactly that. In any case, relativity requires that nonlocalization in space also means nonlocalization in time, so that transitions must not take place instantaneously but must be a gradual process. However, the probability interpretation blurs this aspect of things; it is presumed that a nonlocalized probability distribution can be reconciled with a local and instantaneous process. But once again, we must come back to a criterion for interpretations, that they must be consistent with the mathematical theory. The interpretation is not present in the theory itself. Can point particles really interfere with themselves if they are distributed probabilistically?

Another supposed manifestation of the localized character of electrons is said to be that we can see their tracks in cloud chambers or other detectors. But this is not at all convincing. In the first place, the track must be macroscopic just because it is visible; the fact that it is fairly sharp still allows it a wide microscopic spread. The fact that it forms a fairly smooth line is more a consequence of its high momentum than of its localization.

Schrödinger himself had first wanted to put the "electron wave" on the same footing as the electromagnetic one. However, in the end he was led to concede that his idea would not work. His defeat was based on what he had come to believe were basic characteristics of the theory, which he found, however, to be unsatisfactory. First and foremost, it had become commonplace to speak of quantum phenomena like the drop of the hydrogen electron from a higher to a lower state with the emission of a photon as "quantum leaps," or "quantum jumps," which took place instantaneously. *If* one accepted Born's interpretation, the transition was that from one probability distribution to the other.[12] Although this is conventionally interpreted as a *jump*, which takes place "suddenly" (instantaneously?), the solution of the wave-mechanical problem describes it as a continuous process—once again, though, a continuous transition between probability distributions, which among other things purports to determine the probability that the process takes place at a given time. But one can then ask, When the sudden jump takes place, does the electron become displaced instantly but randomly from one *point* to another, thus violating relativity? When the image of point particles is used, the interpretation does not provide enough information to make a complete and coherent picture. One is reminded of the radioactive decay of an atomic nucleus, which is detected "suddenly" but at an unpredictable time. The situation is more difficult than that, however, because the mixture of initial and final states implies a mixing of the wave functions in the probability distribution, and this mixing is time dependent; in fact, it oscillates with the frequency of the photon being emitted. It is very hard to form a physical picture of what is happening to the particle itself as the probability distribution goes through this strange pattern of behavior. An evident conclusion seems to be, however, that the particle itself may or may not be oscillating, depending on a probability distribution. But if this is the case, what is the mechanism for the radiation that, as a physical fact, will be emitted?

Does it help to visualize this process to consider that the electron is *both* a particle and a wave, which are actually contradictory concepts? Are there not real difficulties in the fact that the probabilities do not depend solely on the amplitudes of the separate states but on their relative phases as well?

What, then, does the Born–Bohr interpretation contribute to an intuitive understanding of the quantum mechanics of this process? Can it really be said to be an "interpretation" at all?

---

[12] The two-state model that we use here as an illustration is an oversimplified representation of reality. Such a model, for example, implies a completely time-reversible process, which does not correspond to reality. Generally, many possible final states will be consistent with the conditions of the problem. This permits us to escape the time-reversal symmetry. However, all of the conceptual difficulties found in the preceding discussion of the two-state model are still present in the refined model.

10

COMPLEMENTARY ATTRIBUTES AND THE UNCERTAINTY PRINCIPLE

Our discussion of the hydrogen atom illustrates a major characteristic of quantum theory, that it does not permit the assignment of arbitrary values to the "classical" dynamical variables. The values of the electron energy are quantized, and so are those of the angular momentum (squared). However, if the value of the component of angular momentum around one axis is specified, those around other axes will be indeterminate, just as are the position and momentum of a free electron. This is reflected in Heisenberg's principle of uncertainty between the values of what Bohr characterized as complementary variables.

Some complementary variables are associated with exclusively quantum properties, like electron spin, which was introduced ad hoc by Goudsmit.

Whatever the interpretation of wave functions may be, the states that they represent have the following characteristics: They are described by a number of compatible variables; they have no internal dynamics (i.e., they describe something that is, not something that is happening); physical change takes place only through transitions between such states; and they have temporal phase coherence, that is, all parts of the wave function act as a single entity, oscillating steadily with the same frequency.

What does it take to describe the quantum state of the electron in the hydrogen atom? So far, we have talked only about its energy; but do the different energy states not differ in some other respects as well? The ground state, at our present level of discussion, is unique, but at the next higher energy allowed, one obtains three equivalent states. They are, in a sense, not really different at all, since they can be obtained from one another by simple rotations in space. These constitute a *triplet* of states. Higher states still appear in multiplets of order $(2l + 1)$, l being any integer. These may be considered as states with angular momentum. Again, the states of each of these multiplets are equivalent and are derivable from each other under appropriate rotations.

What, in fact, comprises a complete description of a quantum state? This is not a question to be answered solely by mathematics. If we solve the Schrödinger equation for the simple hydrogen atom, we find that it contains three parameters, designated as *quantum numbers*. One of these determines the energy; the two others are related to the angular momentum. It appears that the angular momentum, a vector (three-component entity), is not completely specified by quantum mechanics, though it formed, in the primitive Bohr atom, the very basis of quantization. But that was based on the classical orbit picture, in which the orbit was in a plane, and the angular momentum was perpendicular to that plane. When the atom is presented as a wave, that is, a field, its description becomes more complicated. It turns out that all that characterizes the atom, aside from its energy, are one component of the angular momentum and the *total* angular momentum (i.e., one can specify the component of angular momentum in any one arbitrarily chosen direction). Neither of the two other components can then be specified, though their combined contribution to the whole angular momentum can.

The fact that two different components of the angular momentum cannot be specified simultaneously springs from the fact that the determination of one component so changes the atom that the other can be completely indeterminate. Thus, if the angular momentum about one of three coordinate axes is known, the electron will be spinning in such a way that the angular momentum about the other coordinate axes will be in a state of continuous change. In quantum mechanics, the angular momentum is a characteristic of a *steady* state, in which, if one component is known, the two others will not be determined. This matter will be discussed in a more general context in Chapter 15.

Looking at the rationale behind Schrödinger's equation, we can get some feeling for the issues of his atomic model. The wave equations of electromagnetism have some similarity to Schrödinger's equation, except that there are three of them, since the electromagnetic field is a *vector* field; that is, it has three components specifying its value in the three directions in space. In an electromagnetic waveguide, an electromagnetic wave is confined in a "box" of some sort; under these conditions it has a discrete set of allowed frequencies, corresponding to the allowed energy states of the hydrogen atom. These are described by three *quantum numbers*, one corresponding to each direction in space. *Angular momentum* is not one of the wave's characteristics, though it may be for a "free" electromagnetic wave such as that emitted from a radio antenna. In this latter case, just as in the Schrödinger one, two characteristic parameters appear, which, again, are connected to any component of angular momentum of the wave and its total angular momentum. The same is true for an electromagnetic wave confined in a spherical cavity. It appears that this behavior is characteristic of systems with spherical (i.e., rotational) symmetry. The freedom to specify angular momentum in any arbitrarily chosen direction is just the freedom to choose any arbitrary axis of symmetry for the wave. The guiding role of symmetry will appear repeatedly in the discussion of quantum systems; the symmetry is embodied in the wave function.

We must beware of pushing our analogies too far, however. The symmetry axes of electromagnetic waves may be changed by the physical act of rotating the antenna producing it. Is there an atomic analog? It turns out that in fact there is; it is in the act of applying a magnetic field. Just as the spherical symmetry of radiated electromagnetic fields may be broken by giving the generating antenna a definite spatial orientation, so may the symmetry of the hydrogen atom be broken by the magnetic field.

The manner in which this comes about is this: An atom with an angular momentum embodies *a steady* electric current, which makes it effectively a magnet, and this magnet may be turned by an external magnetic field. Note, however, that a steady current does not generate electromagnetic radiation and so does not destabilize the state of the atom; it simply gives it the properties of a stationary magnet.

At this point, however, a special feature of the hydrogen atom appears. Its spectra in a magnetic field were found not to be adequately described by the Schrödinger equation. It appeared that the spatial properties of the magnet were more complicated than anticipated, and this was manifested in the splitting of what the theory predicted to be simple spectral lines. This was interpreted as a manifestation of an *intrinsic* angular momentum of the electron, designated as its "spin." Thus, it appeared that there was a magnetic contribution to the energy that reflected an energy that was due to magnetic interactions. Such an interaction could take place with an external field. But *even in the absence of such a field*, the magnet generated by this "spin" could interact with that due to the angular momentum of the wave function. This is called *spin–orbit* interaction. This intrinsic magnetism of the electron was written into the theory as an empirical hypothesis by S. Goudsmit. Later, however, it was found that such an effect was not arbitrary but required by the laws of relativity. We shall subsequently return to this point.

The moral of this is that experimental fact, and not abstract philosophical speculation, is the ultimate arbiter of theory. Modern particle physics has discovered various new characteristics of fundamental properties that have no classical analogs and, thus, no interpretation in terms of (classically) familiar concepts. This in no way reflects a defect in the theory; it simply highlights our mental difficulties in picturing phenomena in familiar terms.

The consequence for the hydrogen atom is, however, that one new quantum number specifying the spin must be added to describe its state completely. The total angular momentum is again fixed. Its component in any arbitrarily chosen direction can have one of two values, which are plus or minus $h/4\pi$. ("Ordinary" angular momenta come in units of $h/2\pi$.) $h/2\pi$ is generally represented by the symbol $\hbar$.

It is important to understand several features of the wave function. Although what will be said here is in the context of the hydrogen atom, it will also be true of quantum states in general.

First, the wave function has no dynamics; it does not describe something that is happening but something that *is*. It represents a *static* condition of the system, in this case the hydrogen atom. The only information it contains is to be found in the quantum numbers describing it.[13] A consequence of this is that no information can be communicated within the quantum state. Nor can the quantum state describe anything in motion. There is no time sequence of parts of the atom within it.

In fact, the phenomenon of change is not incorporated in the wave function but only in *transitions* between quantum states, these transitions being induced by the action of some external agency on it.

Finally, the wave function is characterized by *phase coherence* of all of its parts; it can only interact as a whole. In this sense, it is indivisible. All parts of it move with the same rhythm. This does not describe evolution in time but is simply a manifestation of the existence of energy. Evolution in time is expressed in terms of transitions between quantum states of different energy and thus different frequency.

---

[13] This statement may seem puzzling, since it appears to involve a spatial distribution of matter, the wave *function*. However, we shall see that this is only derivative information, a consequence of the physical states that the values of the quantum numbers imply. This will become evident when we discuss the Dirac formulation of the theory.

11

An Illustration: Polarized Light

Polarized light provides an example of indeterminacy and of complementary variables. There are two kinds of polarization: linear polarization, characterized by a fixed direction of the electric field of the wave, and circular polarization, which specifies the angular momentum of the light about its direction of propagation.

It is found experimentally that these two types of polarization are not compatible; when one is known, the other is completely indeterminate. A wave that has one characteristic cannot have the other.

In fact, this is true not only of light quanta but also of classical electromagnetic waves. The situation can be easily understood in physical terms, which might lead one to think that quantum mechanics is not so mysterious after all.

One does not usually think of polarized light as essentially a quantum phenomenon. A theory can be developed from Maxwell's equations without resorting to quantizing the electromagnetic field. Nevertheless, we shall demonstrate features that directly parallel those of quantum systems. In this way, for example, we shall come upon the roots of what has come to be known as the *uncertainty principle*, which is due to Heisenberg.

A classical (Maxwellian) electromagnetic wave may be regarded as consisting of a transverse oscillation of electric and magnetic fields at right angles to each other, whose energy is propagated in a direction perpendicular to both of the fields. That there is no field along the direction of propagation springs from the fact that *the photon has no mass* and, thus, no intrinsic inertia, a property that enables electromagnetic energy to be propagated over almost unlimited distance without attenuation. It is possible, by passing it through a properly prepared partially transparent film (of plastic) to arrange that all the light that passes through this film will have its electric field everywhere in a given direction, which we shall call the *x* direction. We shall call this *linear polarization in the x direction*, or, briefly, *x polarization*. If we take the same kind of filter and turn it through 90 degrees, it will let through radiation polarized in the perpendicular (*y*) direction. We will call these filters thus disposed *x* and *y* filters, respectively.

Another feature of light is that it can carry angular momentum around its direction of propagation, in states in which it spins, like a rifle bullet, in either a clockwise or counterclockwise direction about this direction of propagation. We shall call these states, respectively, right circular polarized (rcp) and left circular polarized (lcp). The angular momentum of a single photon is plus $\hbar$ (for rcp) and minus $\hbar$ (for lcp). These are, of course, distinct and mutually exclusive states. Filters can also be produced for each of them.

We can now envisage some experiments with these four kinds of filters.

If light passes through an *x* filter, what remains will not pass through a *y* filter, and vice versa. Let us call these filtering operations P_x and P_y, respectively. Call the *x* and *y* polarized states, respectively, $|x\rangle$ and $|y\rangle$. Let an arbitrary beam of lignt be designated $|a\rangle$. Then, we can say that, in operational algebra,

$$P_x|a\rangle = |x\rangle \qquad \text{and} \qquad P_y|a\rangle = |y\rangle.$$

If the filtering operator for rcp is P_+ and that for lcp is P_-, and the respective states are designated as $|+\rangle$ and $|-\rangle$, then

$$P_+|a\rangle = |+\rangle \qquad \text{and} \qquad P_-|a\rangle = |-\rangle,$$

and, of course, $P_x|y\rangle = P_y|x\rangle = P_+|-\rangle = P_-|+\rangle = 0$.

To go beyond this, we must imagine further experiments. Let us ask, What are $P_x|+\rangle$, $P_x|-\rangle$, $P_y|+\rangle$, $P_y|-\rangle$, $P_+|x\rangle$, $P_+|y\rangle$, $P_-|x\rangle$, and $P_-|y\rangle$? None of these are zero; that is, *some* of the incident light gets through, but not all. The emerging beam in each case has one half of the intensity of the incident one.

Suppose then that we take an arbitrary incident beam and put it first through a + filter and then an *x* filter; we write this $P_xP_+|a\rangle$; that is, the sequence of operations reads from right to left. What will emerge will obviously be a weakened $|x\rangle$ beam. If, on the other hand, we had reversed the order of the filters, what would come through would be a weakened $|+\rangle$ beam. This leads to a queer sort of algebra, such that

$$P_+P_x \quad \text{is not equal to} \quad P_xP_+,$$

and the same is true for all of the other cross-combinations of the two sorts of operators.

The sort of strange consequence that follows from the preceding considerations is illustrated by the following example: Suppose that one starts with some arbitrary beam of light and first uses an *x* filter, thus screening out the *y*-polarized light. If this then impinges on a *y* filter, nothing will get through. In our algebraic scheme, this may be written thus,

$$P_y P_x |a\rangle = 0$$

Suppose now that we interpose a P_+ filter between the two others. In this case, some light definitely does get through. We can test its polarization by using *x* and *y* filters; we discover that it is *y*-polarized. It is found, however, to be only one quarter as intense as the original *x* beam. Still, this seems utterly contrary to common sense, because we had filtered out all of the *y*-polarized light in the very first step, and yet we find some at the end. Or, to put it differently, until we put the last filter in, nothing got through, but when it was inserted, some light did get through. It seems that more filtering, more selecting out, has left us with more light at the end.

Strange though this may seem, it is very typical of quantum processes and is in fact the sort of phenomenon that generates the so-called "uncertainty principle."

Our experiments show that light cannot have a polarization in a fixed direction and a definite angular momentum at the same time. If we had such light, it would have passed through the first two filters intact; the third would then have blocked it all out. This means that measurements of angular momentum and linear polarization interfere with each other; finding the angular momentum destroys the linear polarization and vice versa. They are not mutually compatible properties of light.

Do we then say that these two properties are "uncertain"? It hardly seems the obvious way to describe the situation.

How does the introduction of the quantum affect the preceding arguments? It can best be illustrated by considering again the sequence $P_y P_+ P_x$, where we can now say only that we *may* have some *y*-polarized light at the end. At each step, only one photon can be affected. Thus, after the first encounter with the *x* filter, only a photon polarized in the *x* direction can emerge. Now when this photon encounters the filter for clockwise angular momentum, it may let through only a photon with positive angular momentum. There is actually half a chance that this will happen but an equal probability that the *x*-photon will be reflected back and thus will go no further. Thus, there is only a probability of $\frac{1}{2}$ that a photon will survive this far. However, if a spin-positive photon does emerge, it will now encounter the *y* filter. Again, there is a 50% chance that the photon will be reflected back and so not get through; but if something does get through, it will be one photon polarized in the *y* direction. So while in the previous argument we had concluded that the emerging beam would have a quarter the intensity of the incident one, quantum mechanics tells us that there is only one probability in four that, for each photon entering the system, one will emerge on the other side. In the terms of the conventional quantum philosophers, two processes of "collapse of the wave function" (or quantum state, or whatever) will have occurred. It is really not quite a "collapse" but rather a "transition" or change of quantum state. The photons that did not get through did not vanish but survived in reflected states; the photons are all accounted for.[14] One way or another, conservation of energy must be maintained.

There is much discussion in philosophical circles of *observer-created reality*, which is supposed to establish our interconnectedness with the physical world. Thus, if we put a polarization filter in the path of a photon to measure its polarization, we are *creating* what we are attempting to measure. Of course, we set up our measuring instruments (or we make our interventions in the physical world) precisely for that purpose. But the physical world does not care, so to speak, whether we act by intent or not. We do affect the physical world, and it affects us, but there is no evidence that our interactions are not subject to physical laws that would be the same if we were not here at all. There is surely nothing either mysterious or mystical about it.

Why can we not know the angular momentum of a photon of radiation and also the direction of its polarization? Different experiments show that both of these properties can exist in photons. The point is, however, that a photon that has one of these properties cannot have the other. This is an ontological reality. The properties cannot conceivably coexist. However, there is a physical reason for this. The angular momentum of a photon is associated with the rotation of the electric vector about the direction of propagation. If the electric field is rotating, it cannot have a fixed value. The situation can also be approached from the viewpoint of symmetry. Rotations can either be clockwise or counterclockwise about a given direction. In an electromagnetic wave, the only things that can rotate are the electric and magnetic fields, which rotate together, keeping a right angle between them. But, of course, if the electric vector were fixed, it would be symmetric with respect to the two directions of rotation. In that situation, angular momentum cannot exist.

In the same sense, something cannot be a point that has zero dimensions and a wave that is spread out through space. In quantum mechanics, the momentum of a wave is derived from its wavelength, which defines its translation properties in space. For a single direction, there are two opposite directions of translation. A point sits between the two; it does not distinguish between the two directions. Whereas angular momentum implies a breaking of rotational symmetry, momentum implies a breaking of translational symmetry.

So the problem is not one of uncertainty at all, in the sense in which the term is usually understood. Bohr came closer to reality in defining pairs of properties like position and momentum, electric field direction, and angular momentum as *complementary*, each affording one way of assigning attributes to a physical system.

---

[14] There is, of course, a possibility that a photon that did not penetrate a filter could be absorbed in it rather than being reflected, but that does not change the essential features of the argument.

What then is the meaning of the "uncertainty principle"? The question it answers is this: Suppose that one does not know the direction of the electric field of a photon precisely but only within certain limits. One may ask, If one then measures its angular momentum, what are the limits on the value that might be obtained? The values of the angle that the field might make with a fixed direction have a range of π. The limits of the angular momentum that might be obtained are $\hbar$ and $-\hbar$, a range of $2\hbar$. The product of these "uncertainties" is then $2\hbar$ (i.e., one can specify the component of angular momentum in any one arbitrarily chosen direction). Of course, it is not accidental that the product of these complementary uncertainties is proportional to Planck's constant. In fact, it is possible to prove a very general theorem: that the product of the uncertainties of complementary variables must be not less than $\frac{1}{2}\hbar$.

12

Heisenberg and Measurement

Werner Heisenberg arrived at quantum mechanics by a quite different route. For him, the essence of the theory was in the fact that the process of measurement of one dynamical variable of a system (the position of a particle, for instance) caused a disturbance of the system that rendered the determination of a complementary variable (in this case momentum) uncertain. Heisenberg thus started with the concept of a point particle and considered the resulting "uncertainty" to be a consequence of the measurement process. In this way, he rejected the classical goal of describing the behavior of physical systems and made the theory into one in which the central goal was to determine the results of experiments. It was through this approach that the human observer came to be a central actor in the theory. Whereas Schrödinger's waves suggested a physical reality, Heisenberg's theory was concerned not with the physical world itself but with how we came to know it.

At the beginning, two different viewpoints were put forward for quantum phenomena. We have spoken primarily about the Schrödinger one, which originated from the "particle wave" hypothesis of de Broglie. The other, which is due to Heisenberg, is much more convoluted and is difficult to explain in visual terms because it leans heavily on mathematical formalism and in fact explicitly precludes, as stated by Heisenberg himself, "a visual description of the atom." His ideas take as their starting point, however, the idea of the trajectory of a particle. From the very beginning, he rejects the idea of the objective reality of physical processes in favor of the hypothesis that only concepts used to describe physical experiments have a place in the structure of the theory. In this way, he places the problem of measurement at the center of the theoretical structure.

It is interesting to note the consequences of this hypothesis, since it also assumes the human actor, the investigating physicist, to play a central role in physical theory. By this process, "objective reality" is discarded, and the door is opened to subjectivism. This represents a turning point in the history of physics, if not of science itself, with as a consequence the continuing widespread concern with the supposed paradoxes and philosophical difficulties of quantum theory, despite its now long history of spectacular successes. The phenomenon seems to parallel closely the unfortunate experience of the ether theory of the late 19th century and to illustrate the hazards of building a metaphysical superstructure around physical theory that is not intrinsic to the structure itself. To put it simply, the human experimenter does not appear in the equations.

To try to gain a better understanding of Heisenberg's thinking, let us indicate the general lines of his argument.

First, Heisenberg imagines that an electron (say) has a trajectory. He then asks how one could attempt to observe it experimentally. Since the trajectory proceeds continuously point by point, one would have to measure its position point by point. The tool to be used is a beam of light, which will reflect off the moving particle. To measure even a tiny portion of the path, one would have to use light of a very short wavelength. But quantum mechanics tells us that the shorter the wavelength, the higher the energy of its quanta, the photons. But to strike a particle with such a high-energy object would eject it violently from its orbit, thus aborting the process of following the path. This illustrates, he says, "the need to relinquish the concept of an electron path *and to forego a visual description of the atom*" (my emphasis). The first half of this statement is unarguable; the second is less a consequence of the first than a confession of limits to the imagination. It obviously simply begs the question in assuming that the only conceivable visual description is one in terms of classical trajectories!

Heisenberg then, still without abandoning the trajectory idea, evokes a mathematical device not intuitively evident to the noninitiated: that the trajectory can be analyzed in terms of wavelength or frequency components, which, in classical theory, would determine the radiation emitted by the classical electron. From observation of this pattern of frequencies, the path could be reconstructed. But, of course, spectroscopy has shown us that the radiation emitted by an accelerating electron does not have a continuous spectrum but one consisting of a set of waves with discrete frequencies. That is to say, only certain components of the path will remain in quantum mechanics. Each of these components will correspond to discrete transitions between one energy state of the electron and another. Each of these transitions will, in turn, be determined by parameters characterizing the two states between which the transition takes place. The complete set of such parameters constitutes what mathematicians call *a matrix*, so Heisenberg's theory came to be known as *matrix mechanics*.

It is interesting to note, however, that though Heisenberg's thought springs from the trajectory model, the idea of discrete frequencies does not require this setting but can stand on its own. Thus, the link with trajectories is not in itself logical. In fact, if one discards all components of the path that do not correspond to allowed spectral frequencies, one is left with not a trajectory but a wave. This is in fact the very essence of waves—to be characterized by wavelengths and frequencies.

It is perhaps not surprising, then, that Schrödinger was able to show that, despite appearances, Heisenberg's matrix mechanics and his wave mechanics were essentially equivalent. One could arrive at quantum mechanics by two routes formally and philosophically different but nevertheless equivalent.

13

More on Measurement

Philosophical discussion of the "meaning" of quantum theory has over the years reflected the philosophical differences between the Heisenberg and Schrödinger viewpoints. A long-standing dispute took place between Niels Bohr and Albert Einstein, with Bohr pursuing the philosophical approach of Heisenberg and Einstein adhering to what came to be known as the "realistic" view.

Einstein put forward two principles that he thought should govern any valid physical theory. One was "realism," the contention that properties that could be measured with certainty were physically "real." The other was "locality," that if two "systems" were in dynamical isolation from each other, a measurement on one could not affect the other. John Bell was able to show that these hypotheses alone put limits on the results of certain measurements. He showed, and it was experimentally verified, that these limits could be exceeded in quantum mechanics.

Shortly before his death, Bell shocked the physical world again by contending that a valid theory should "stand on its own feet" and not depend on subsidiary hypotheses not intrinsic in the theory itself, such as the introduction into its formulation of the role of measurement. This chapter deals with how this led to the old arguments flaring up again.

Not long before his death, John Bell, best known for his Bell's inequality, which established that quantum mechanics was not a local realistic theory in the sense of Einstein, published an article in *Physics World*[15] raising profound questions about the role of interpretation in quantum theory. He says,

---

[15] August 1990, pp. 33–40.

Surely, after 62 years, we should have an exact formulation of some serious part of quantum mechanics. By "exact" I do not of course mean "exactly true." I only mean that the theory should be fully formulated in mathematical terms with nothing left to the discretion of the theoretical physicist.... [Further,] however legitimate and necessary in application [the following words] have no place in a formulation with any pretention to physical precision: *system, apparatus, environment, microscopic, macroscopic, reversible, irreversible, observable, information, measurement.*... [On] this list of bad words... the worst of all is "measurement."

Note that this is not an injunction against the use of these words in discussions of quantum mechanics, only against their incorporation as essential elements of the theory itself. As Bell himself notes, such language should be confined to the *applications* of quantum mechanics only.[16]

Bell continues on a more sardonic note:

It would seem that the theory is exclusively concerned about "results of measurement" and has nothing to say about anything else. What exactly qualifies some physical system to play the role of measurer? Was the wave function of the world waiting to jump for billions of years until a single-celled living creature appeared? Or did it have to wait a little longer, for some better qualified system... with a Ph.D.? If the theory is to apply to anything but highly idealized laboratory operations, are we not obliged to admit that more or less "measurement-like" processes are going on all the time, more or less everywhere? Do we have jumping all the time?

In 1991, *Physics World* became involved in a controversy over J.S. Bell's critique of quantum mechanics, which drew responses from, among others, R. Peierls and K. Gottfried. Apparently, a "Bell Memorial" volume is on the way which will have a collection of articles on the subject. The gist of the matter is that Bell argues for a formulation of quantum mechanics that foregoes all use of terms that do not appear in the mathematical structure of the theory.

Van Kampen in this issue takes another tack. He defines quantum mechanics as a *macroscopic* theory that "works." That is, it gives the right answers to experiments. His "theorem 4" contends that "whoever endows [the wave function] with more meaning than is needed for computing observable phenomena is responsible for the consequences"—with the apparent implication that they will be dire. He does deal properly with incoherent macroscopic systems (e.g., Schrödinger's Cat) in that he identifies the incoherence with loss

---

[16] Peierls would make an exception of the word *information*, on the grounds that the subject of the theory *is* precisely information about the physical world rather than the nature of the physical world itself. While the theory can undoubtedly be interpreted in this way, the question is whether it *must* be so considered, and if not, why quantum mechanics should be considered as a fundamentally different kind of theory from all the others used by physicists. Can one really distinguish two kinds of theory with quite different goals? Or is it to be contended that *all* theories should be interpreted in this way?

of phase information, so that predictions are reduced to probabilities only. But he seems to be satisfied with that, and he makes no mention of coherent macroscopic systems or states. All quantum mechanics, for him, is then in density matrices (appropriate to ensembles). Again, individual particle quantum states as physical entities seem to be abolished from his picture. When it comes to macroscopic quantum systems, we shall subsequently present arguments showing that this is correct.

Peierls, in his book *Surprises in Theoretical Physics* and elsewhere,[17] has expressed what he claims to be the real essence of Bohr's Copenhagen interpretation, which may be summed up in the statement that "quantum mechanics does not tell us about what the physical world really is, but only what we can know about it." He then treats several problems from the viewpoint of gain and loss of information at different phases of phenomena. This proposition makes it appear as though the central issue of the theory, unlike that of any that has preceded it, is *our knowledge* of the physical world and not that world itself. This subjectivist view is, of course, nothing new but was not widely held until revived by Bohr. In any case, it raises the question, Is there any difference between what we can know and what there is to be known? If what we can know is sufficient to give a complete picture of the physical world, what is the issue in the distinction between knowledge and "reality"? I have used the word *picture* rather than *description*, implying a set of images (see Faraday) rather than a recital of numerical properties. An interpretation is concerned specifically with such pictures.

Finally, Gottfried defends the traditional view. His is the most comprehensive critique of Bell's position. His carefully argued case rejects Peierls' "orthodoxy" but retains the notion of the central position of *measurement*, as does Peierls. He specifically starts, however, from the assumption that Schrödinger's equation should be interpreted in terms of point particles.

It must be admitted that in noncoherent macroscopic systems wave functions play no role (I would say do not exist, though to the best of my knowledge that contention is not widely accepted). Thus, density matrices and "mean values" dominate; statistical mechanics is the right vehicle of analysis. In this, van Kampen and Gottfried are apparently in accord. But it is not enough to deal with fields only through their amplitudes; phases, as Feynman emphasized, are absolutely central in elemental systems. Anyway, that is where all the paradoxes dwell, and my original concern was with paradoxes.

Like Bell, I do not believe that measurement should have any basic role in quantum theory. All of the work being done seems to point to the conclusion that quantum mechanics itself is the basis for analyses of the measurement problem, which brings us back to the thesis that measurement is a quantum process and not an integral part of the theory itself.

---

[17] In Defence of "Measurement," Phys. World, January 1991. *Surprises in Theoretical Physics*, p. 32, Princeton University Press, Princeton, New Jersey, 1979.

It is interesting to go back to electromagnetic theory and raise the problem of interpretation; the basic question is, What fundamental difference is there between electromagnetic theory and quantum mechanics in this regard? In what way are the electromagnetic fields fundamentally different from wave functions representing electron fields? **In terms of what can one "interpret" them?** How does one know that they exist as physical entities rather than as elements of a mathematical structure that "gives the right answers" or that simply represents the sum of our knowledge of electromagnetic phenomena? Would van Kampen dismiss them from their ontological pedestal and regard them as having no significance except as devices for computing observable phenomena, for instance, the electromagnetic behavior of matter?

There does not seem any cogent reason to find in these issues a fundamental flaw in quantum mechanics itself. But Bell's thesis that the essence of the theory should not depend on hypotheses about measurement, and that its incorporation into the body of the theory clouds consideration of the real meaning of the theory, could well be of even more import than his by now famous "inequalities." Certainly, it is a fresh new idea in a field that has generated continuing controversy for decades.

14

FURTHER REFLECTIONS ON THE BOHR–PEIERLS INTERPRETATION

The arguments over interpretation of quantum mechanics largely take place outside the framework of the theory itself. The questions raised are then not technical ones but reflect a wider philosophical framework. For most of them, no authoritative answers exist. We attempt, in this chapter, to identify some of the areas of dispute. Can physics explain why the laws of physics are as they are? Is there a difference between what the physical universe is "really like" and "how it works" as described by the laws of nature? Are these even meaningful questions? What is the relationship between the theoretical constructs of our minds and the realities of the physical world? An effort is made to relate these questions to the nature of the scientific enterprise.

Let us return to the question, What is the difference between *what the theory really is* and *how it really works*? Feynman has some difficulty with the same problem.[18] "I think that I can safely say that nobody understands quantum mechanics." As an alternative, he says "I am going to tell you what nature behaves like." If, he says, one asks the question, "But how can it be like that?" the answer is, "Nobody knows how it can be like that."

The problem with questions like, What is the universe *really* like? and How can it be like that? is that they have no context. That is to say, they pose the question, How can one conceive of a basis for answering such a question? In terms of what can it be answered? It is somewhat similar to the question, Why

---

[18] *The Character of Physical Law*, p. 129, M.I.T. Press, Cambridge, MA, 1965.

is a law of physics what it is?, or How was the universe created? It also calls to mind Newton's absolute space and absolute time, which provided the framework of a stage on which the story of nature could unfold. The only imaginable answers in fact lie outside the scope of science—in divine revelation, for example. It may be asked, Why are the laws of physics as they are?, leaving no laws on which to lean for answers.

St. Augustine grappled with such problems in a religious context but, having a questioning mind, he found not answers but further questions. "What was there before the Creation?", he asked, and his answer, which can be paraphrased as "Before the creation there was nothing (but God?) and so there was no time, so the question is without meaning," is one that can withstand criticism even in this scientific age.

It may be thought by many that the purpose of science is to "explain the world" but, despite the pretensions of some physicists of more arrogance than wisdom, that does not comprise explaining how the universe came about or why it is the way it is. A more fruitful and realistic question, the one Feynman addressed in professing his inability to cope with the other one, is, How does it work? The question is, Is this really a different question? Is there any way to answer the question, What is it? other than with How does it work?

Consider, for example, the laws of physics. On close analysis, it is apparent that their status is that of principles that relate many phenomena to each other. Thus, Newton's law of gravitation related the fall of an apple, the flow of the tides, and the orbits of the planets. It does not say *why* gravity is but *how* it makes the world as it is comprehensible. In the same way, the electromagnetic theory of Faraday and Maxwell did not explains *why* there is electricity and magnetism but *how* they relate to each other, to light and color, to radio and x-rays and microwaves, to gamma radiation, and to lasers. Were there not laws of nature, the world would be infinitely complicated and frightening; with laws, complexity is reduced to simplicity. Thus, the great goal of modern physics is defined as that of finding a "grand unified theory," which would constitute a basis for understanding how all physical phenomena relate to each other. That is what leads us nowadays to try to relate the physics of the astrophysical phenomena to particle physics, the physics of the ultimate components of matter, if such there be.

Is this not an answer to the question, What is the nature of the world of our experience? Is any other sort of answer conceivable within the framework of science?

The response that comes to mind, however, is this: All of our experience tells us that our understanding of the laws and relationships of the natural world is only partial and constantly changing. Some, like Stephen Hawking, may think that "final answers" are close at hand, but closer examination reveals that a seemingly inexhaustible series of basic questions still remain to be answered. On the other hand, it surely cannot be true that the true nature of the world, the ultimate realities, are undergoing constant change and refinement! Thus, our "reality" is confined by the limitations of our understanding. I suppose that in that sense physics is about "what we can know about the world." But note that I do not say that as something that distinguishes quantum mechanics from our other theories. The proposition is universal and recognizes no special status for any particular theory. This is my ultimate point of contention: the thesis that quantum mechanics is a theory like any other. In particular, it does not shed any light on the relation of humans to the physical world, though this is the contention of some scientific philosophers. It does not give a special status to the process of humans experimenting with physical phenomena, that is, with the process of *measurement*; rather, it is a theory of the nature of the physical world itself.

A by-product of this thesis is that it assumes that, incomplete though our knowledge and understanding of the physical world are, the object of the understanding is real and external to us; we are part of it and it not a part of us. In short, it assumes that there is an underlying reality to be understood. This is clearly not a self-evident truth. But to deny this implies a denial of the adequacy of science itself to realize its own goals. There is no way in which we can resolve this question, or even to put it to critical test, for if there is something about the physical world that it is in principle impossible to incorporate in our structure of understanding, not only must it be outside our understanding but also outside our reality. It must occupy a different space than that explored by our science and is thus irrelevant to science.

It must be admitted that elusive questions remain which may confound us further. For instance, can the human mind in principle understand itself? The answer may well be negative; that is, the mind may not simply be integrated into scientific laws. If it is, it must be so many levels removed from our present competence as to reduce the question to unfettered speculation.

All that remains to us is to continue to work within our present confines, enlarging step by step the domain of our imperfect and incomplete comprehension; in humility, being ever more conscious of what we do not know than of what we do. At this point, it does not really matter; if there are things we *can* never know, we can also never know whether they do or do not exist.

15

Toward the Dirac Viewpoint

Paul Adrien Maurice Dirac, one of the greatest physicists of modern times, set forth a formulation of quantum theory that encompassed the versions of both Schrödinger and Heisenberg as different representations of a general mathematical structure, thus overriding the special philosophical biases of each. His theory encompassed in mathematical form the major features of the quantum world: the existence of quantum states, the complementary properties of quantum particles, and the significance of the noncommutativity of dynamical operators. And most importantly, he showed how to determine the properties of the operators representing the dynamical variables in terms of the classical relations between them.

In principle, and in numerous specific instances, he showed that quantum states and quantum transitions could be determined from the "commutation rules" of the operators alone, which themselves determined the "uncertainty relations" between them.

To remove a bit of the mystery of noncommuting operators, we show, as a simple example, that the operators for rotation of objects about different directions do not commute with each other. In a separate section, we present, for the mathematically minded, a thumbnail sketch of the Dirac linear operator scheme.

Finally, we note that the "uncertainty relations" are not about our ignorance, as claimed; rather, complete information about quantum systems can be deduced from them.

The next monumental step in the development of quantum mechanics was made by P.A.M. Dirac, who found a mathematical formulation of quantum theory that did not depend either on the wave or the trajectory viewpoint but incorporated both as special cases. In the process, he was able to extract the essential constituent of the theory, of which both Schrödinger's and

Heisenberg's theories were realizations. The common thread was the mathematical discipline of linear algebra. This is effectively an algebra of operators. The basic elements of the scheme are quantum states of physical systems. These states can be described in terms of dynamical variables (such as energy, momentum, and angular momentum), which characterize the physical properties of physical entities. *Linear* signifies that "effect is proportional to cause." It is convenient to use a shorthand to express these relationships. The states to be acted on will be designated by symbols like $|a\rangle$ and will be characterized by the values of the set of dynamical variables needed to describe them completely. Only experience can tell us what these are, as we shall see when we discuss the states of atoms. The mathematics must then be tailored to that experience.

What does it mean physically when we operate on a state with an operator associated with a physical property? It means that we are searching to discover the status of the dynamical variable that the operator represents in the system represented by the state. The result of the operation will cause a transition of that state to other states; each has a matrix element. That matrix element has an amplitude whose square represents the probability of the transition to the particular resulting state. It also has an amplitude, which represents the phase change associated with the transition. It is this phase that is responsible for all interference phenomena. Together, these constitute the matrix element of the transition. These matrix elements carry all the information necessary to describe the dynamics of quantum processes, that is, the way in which quantum systems vary with time. This is how the description of physical processes proceeds in quantum theory, rather than by the following of the point particles whose trajectories are studied in classical Newtonian mechanics.

What if the numerical value of the operator is known for the state operated on? In that case, the operator does not change the system and can be replaced by the value of the dynamical variable it represents. Such states are known as *eigenstates* (proper states) of the dynamical variable. For instance, in the eigenstates of the electron in the hydrogen atom, all of the quantities necessary to give each state a full description have definite values. These are most often discrete values, which define the quanta of the system. It does not matter in what order the operators are applied, since in each case the state is not changed. This is, of course, another way of saying that the different properties of the atom are compatible.

All of this can be summed up in a simple algebraic system, which, for those who have a taste for technical precision or can admire the elegance of the language of mathematics, is briefly summarized in Section 15.1 at the end of this chapter.

In Chapter 11, we illustrated, by the example of polarized light, different ways of characterizing the quantum states of photons. This revealed that certain properties were not compatible (by the very physical nature of the photon). The incompatibility, we showed, manifested itself in the fact that the operators for two incompatible dynamical variables did not commute; that is, we arrived at different final states when we reversed the order of the operators. There are simple examples of this sort of thing in ordinary experience. Polarized photons provided one such example; another, much more simple conceptually, concerns rotations of objects.

Imagine a rectangular box. Choose one corner for a base point, and think of the three edges radiating from it. (They form a Cartesian coordinate system.) Label them x, y, and z. (Figure 2a). Now rotate them clockwise through 90 degrees about the x axis to get the configuration in Figure 2b. Then rotate again through 90 degrees about the y axis to arrive at the configuration in Figure 2c.

Return to the original state (Figure 2d), and this time rotate first about the y axis and then about the x axis. The first operation leads to Figure 2e, and the following one leads from Figure 2e to 2f. What we have is very different from Figure 2c, so that the two operations do not commute. But just as in the case of polarized light, it is possible to see why this is so. The y axis in the first case is in the position to which it had been rotated by the x rotation, while that in the second case was in its original position. The second operation is on the *result* of the first. This is a new kind if symbolic algebra, different from that of ordinary numbers but not at all uncommon either in the pure mathematics or its applications to physics.

A further mathematical theorem is quite easily proven (though we will not do it here). It says that from the commutator $K_1K_2 - K_2K_1$ it is possible to deduce how the spread of possible values of one dynamical variable will limit the spread of values of the other. In quantum mechanics, these spreads are called (rather inappropriately) *uncertainties*, though this is not the usual meaning of the word. They do not imply a lack of knowledge but an intrinsic incompatibility of the concepts associated with the physical variable. A simple example is the wave–particle duality that is sometimes used to describe quantum theory.

One interesting consequence of the theorem is this: It often turns out that the eigenvalues of operators, the possible values in quantum states, can be deduced simply from the commutation relations. In fact, in principle they always can, though it is sometimes difficult. Thus, both wave functions and Heisenberg matrices are dispensable.

There is something quite striking here, which cannot be too strongly emphasized: The uncertainty relations have always been portrayed as signifying that some features of quantum systems are unknowable; that they indicate, so to speak, that nature is hiding something from us. It has seldom been recognized that, on the other hand, *they determine everything that is knowable*, that they give us a complete characterization of the properties and behavior of physical systems. What we *cannot* specify is not a result of incomplete knowledge; what we cannot know *does not exist*. This is an ontological statement; waves do not have position, electromagnetic waves with their electric fields in a fixed direction cannot have angular momentum.

So the uncertainty relations are not negative but positive; they are the key to a complete understanding of physical reality.

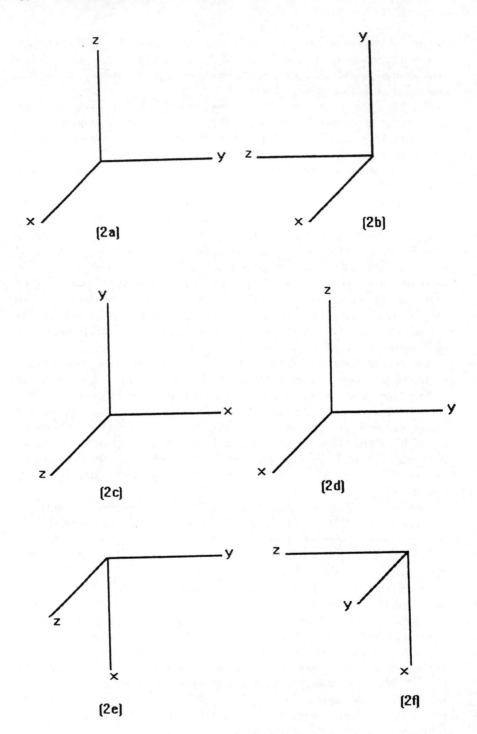

This seems to be a very abstract formulation, which avoids even the question of variation in time. It leads us to ask where the physical content of the Dirac scheme is to be found. The answer is, of course, in the form of the operators K. But, as already remarked, Dirac has given a very precise rule for constructing the properties of the quantum operators from classical dynamical variables. That there can be more than one explicit representation of the operators having these properties, or if they are simply left as abstract, is no cause for alarm. It is not even clear that space and time coordinates must be incorporated into the scheme. In the Schrödinger representation, which uses differential operators, however, they do appear. This is not an accident. If we take the coordinates, collectively designated x, as dynamical variables, we can form products such as $\langle x|n \rangle$ or $\langle n|x \rangle$. This assumes that the possible values of x as an operator are the numerical values of the coordinates. In any case, the products are functions of the coordinate x and are in fact the Schrödinger wave functions for the states $|n\rangle$, ($\langle x|n\rangle$, their complex conjugates $\langle n|x\rangle$). These must be complex numbers, since they must have both amplitude and phase. It is required by their wavelike character.

But is the position, as specified by the coordinates, really a dynamical variable, especially since a particle in quantum mechanics may not have a precise position? In fact, the question is not really important; it does not matter whether it is treated as a dynamical variable or not. We are here concerned only with representations of the operators. Physical results are actually independent of representation, so the specific representation used has no special physical significance. All that is required is that the representative "states" form a complete set. We make the plausible assumption that this is true for coordinate states.

We now see, then, that the Schrödinger scheme is just the expression of Dirac's scheme in a spatial representation (space–time in relativity). Does this mean that the wave functions are then physically real? If the concepts of space and time are taken to be fundamental to our vision of the physical world, the answer must be positive.

15.1. THE DIRAC VERSION

The Dirac formalism implies that there are quantum states, mathematically described by what Dirac calls *kets* $|n\rangle$, which are operated upon by the operators of dynamical variables K. If the operation of K on $|n\rangle$ reproduces $|n\rangle$ to

Figure 2.
(a) The original configuration. (b) Rotated through 90 degrees about the x-axis. (c) Rotated about the y-axis. (d) The original configuration. (e) Rotated 90 degrees about the y-axis. (f) Rotated 90 degrees about the x-axis. Compare (c) and (f).

within a factor; that is, if $K|n\rangle = k_n|n\rangle$, this means that the system is not disturbed by the operation, so that is definitely determined to have a precise value of the variable K, namely k_n. If, on the other hand, $K|n\rangle$, produces another state $|n'\rangle$, that is,

$$K|n\rangle = \mu|n'\rangle,$$

the dynamical variable K does not have a definite value in state $|n\rangle$ but creates another state $|n'\rangle$. Another way of saying this is that K brings about a *transition* from state $|n\rangle$ to state $|n'\rangle$.

Dirac defines a complete set of *bra* states $\langle n|$ associated with their respective ket states in such a way that $\langle n|n\rangle = 1$ (this is called the *normalization condition*) and $\langle n'|n\rangle = 0$ (orthogonality). On the other hand, if we multiply the preceding equation on the left by $\langle n'|$, we have

$$\langle n'|K|n\rangle = \mu\langle n'|n'\rangle = \mu;$$

this is called the *matrix element of the transition* from state $|n\rangle$ to state $|n'\rangle$. These are precisely the *matrix elements* of Heisenberg, stripped now of all association with trajectories. By the same token, $\langle n|K|n\rangle = k_n$; these *diagonal matrix elements* are the values of the dynamical variable K in the states in which it is well defined.

If $|n\rangle$ happens to be a state in which two variables K_1 and K_2 are both well defined, with numerical values k_{1n} and k_{2n}, respectively, then

$$(K_1 K_2 - K_2 K_1|n\rangle = (k_{2n}k_{1n} - k_{1n}k_{2n})|n\rangle = 0.$$

This says that for compatible dynamical variables, the commutator is zero; or, alternatively, if the commutator is zero, the variables are compatible. This latter statement is valid because, if

$$K_1|n\rangle = k_{1n}|n\rangle,$$

then $(K_1 K_2 - K_2 K_1)|n\rangle = 0$ or $K_1(K_2|n\rangle) = k_{1n}(K_2|n\rangle)$ is also a state for which K_1 is determined with definite value k_{1n}.

16

FEYNMAN'S PATH INTEGRAL METHOD[19]

Richard Feynman, who seemed always to be able to create his own physical world, produced an alternative version of quantum theory from the conventional one. Although it is formally quite different from the conventional one, it is completely consistent with it. What it does do is conjure up a different set of images, which open up interesting new insights. Nowhere does it invoke probabilities, nor does it suggest collapse of the wave function on measurement. Rather, it reaches out to encompass all space. It talks not about particles but about paths, which carry phase from point to point. We illustrate it with a familiar example from geometrical optics.

Richard Feynman had a very original mind, and, finding himself uncomfortable with the standard interpretations of quantum mechanics, he reinvented the subject in a manner more to his liking. What is astonishing is that, although his approach looks quite different from the conventional one, it can be shown that it leads to exactly the same results.

His approach is simple enough in concept but involves new mathematical techniques that can become quite complicated. On the other hand, it lends itself to the treatment of problems that are very difficult to deal with by standard methods. The method can thus be justified on practical grounds alone, but alternative ways of viewing problems also can often give new

---

[19] The subject is fully explained in Feynman's book *Q.E.D.* (Princeton University Press, Princeton, NJ, 1985). Our purpose is merely to give an illustration of the ideas behind it.

insights; Feynman always insisted that it was important to "take the world from another point of view."

The idea of the theory is perhaps best illustrated by the classical analog of geometrical optics. One of the theorems of geometrical optics is that if light emitted at a point A is reflected at a mirror and viewed at point B, it must follow a path such that the angle of the incident beam to the mirror is the same as that of the reflected beam. This is known as *Snell's law*. Now, geometrical optics is a strange subject precisely because it speaks of *beams* of light as if they followed trajectories. On the other hand, we know that light is in effect a wave phenomenon. *Geometrical* optics, a subject that William Rowan Hamilton developed in a particularly general and lucid way, is really optics at *microscopic* wavelengths, that is to say, the wavelength of the light is short compared with the size of the macroscopic elements in the optical system being considered. The higher the momentum of the light in the direction of propagation, the greater will be the transverse momentum due to its spreading. This in turn, by the uncertainty principle (which applies basically to all waves), keeps the spatial spread of the beam narrow. In this way, one can have narrow beams of high-energy, and thus high-frequency, light.

Feynman's interpretation of Snell's law is, however, quite different and depends on consideration of the phase of the light along its path and thus on interference. Feynman's hypothesis is that light can in fact get from A to B by *any path whatsoever*, with equal probability. However, along any path, its phase will change in a prescribed way, depending on changes of a classical dynamical variable called the *action*. Therefore, if light goes by different paths which differ in length, the phase with which it arrives at B will in each case be different. Thus, there will be great deal of destructive interference, and the signal received at B when averaged out will be negligible. There is, however, an exception. It is possible to show that if one takes the point on the mirror where Snell's law holds, immediately neighboring paths will have almost the same length from A to B. That is, there will be, at and in the near neighborhood of this point, a *constructive* interference that will enhance the light intensity along that path. Thus, Feynman's prescription leads directly to Snell's law, by virtue of the interference of contributions from the diverse paths. Feynman's idea was then to use the same point of view for quantum waves.

One thing is clear—there is no room for the probability interpretation in this treatment. Quite the contrary, in fact, since its argument is based on phases, which have no place in that interpretation. It is important to be clear about this, because it is tempting at first sight to think that Feynman has reintroduced *trajectories*. But it is essential to remember that a *single* path has no physical connotations in Feynman's scheme. Single point particles cannot carry phase from point to point. The correct physical phenomena are a function of the interaction of *all* the paths.

Recently, Stephen Hawking has used Feynman's technique in another context. He invokes something that he calls the "wave function of the universe," which presumably is meant to be a complete quantum description of the universe. In fact, he states specifically that it describes "the universe and everything in it." But this cannot be what is really meant, because the universe contains an almost infinite amount of information, and his wave function can actually only be the wave function of a simple mathematical model. But he claims that each Feynman path represents something physical—a complete description of a possible history of the universe. On the other hand, even the word *history* is suspect, since evolution takes place in something called "imaginary time." Thus, his work is revealed as a mathematical exercise without real physical context. Feynman's paths are not the physical paths of real objects[20]; reality is described only by adding together the contributions of an infinite number of these imaginary paths, which are individually simply mathematical constructs. That the underlying physical reality is one of *fields* and not point particles is clear, for the method does not involve considering one path *or* another but rather requires considering *all paths at once*. There is a world of difference between the realities of "this or that" and "this and that."

One can only applaud Feynman's ingenuity in developing this original method, which, if correctly used, plays a very useful role in physics. What is more dubious is to try to read into its mathematical tools a physical meaning of their own. Feynman did not restore trajectories to quantum physics; he simply discovered an alternative mathematical scheme that reproduced all the features of the conventional theory. One can be certain that he had no illusions of using it to create the ultimate theory of the physical universe.

---

[20] Feynman himself has affirmed this. See R.P. Crease and C.C. Mann, *The Second Creation*, p. 138, Collier-Macmillan, New York, 1986.

17

ARE FIELDS ALL?

Nobel Prize physicist Steven Weinberg gives us a picture of a world constituted solely of interacting quantized fields, in which particles are "reduced to mere epiphenomena."

Steven Weinberg presents us with a picture of the physical world as nothing but a set of interacting quantized fields; it is a picture that seems to leave no room for purely nonphysical or conceptual entities such as probabilities in the quantum-theoretical view of nature. It may be seen as a rejection of the conceptional dualism of particles and fields:

In 1926 Born, Heisenberg and Jordan turned their attention to the electromagnetic field in empty space and . . . were able to show that the energy of each mode of oscillation of an electromagnetic field is quantized. . . . Thus the application of quantum mechanics to the electromagnetic field had at last put Einstein's idea of the photon on a firm mathematical foundation. . . . However, the world was still conceived to be composed of two very different ingredients—particles and fields—which were both to be described in terms of quantum mechanics, but in very different ways. Material particles, like electrons and protons, were conceived to be eternal. . . . On the other hand, photons were supposed to be merely a manifestation of an underlying entity, the quantized electromagnetic field, and could be freely created and destroyed. It was not long before a way was found out of this distasteful dualism, toward a truly unified view of nature. The essential steps were taken in a 1928 paper of Jordan and Eugene Wigner, and then in a pair of long papers in 1929–30 by Heisenberg and Pauli. They showed that material particles could be understood as the quanta of various fields, in the same way as the photon is the quantum of the electromagnetic field. There was supposed to be one field for each type of elementary particle. Thus, the inhabitants of the universe were conceived to be a set of fields—an electron field, a proton field, an electromagnetic field—and particles were reduced to mere epiphenomena. In its essentials, this point of view has survived to the present day, and forms the central dogma of quantum field theory: *the essential reality is a set of fields* subject to the rules of special relativity and quantum mechanics; all else is derived as a consequence of the quantum dynamics of those field.[21]

A currently fashionable idea is that, just as electron wave functions represent probabilities of the location of a point particle rather than physical fields, Maxwell's equations should play a similar role for photon "particles."[22] For example, Pagels, after quoting Weinberg, says, "What is meant by *quantizing a field* is analyzing a field like an electromagnetic wave in terms of its associated quanta, the photons. The intensity of the electromagnetic field at a point in space gives us the odds of finding a photon there." But in fact "analyzing a field like an electron wave in terms of its quanta" does not yield point particles but quantized values of field energy. It is therefore difficult to interpret what is meant by "there" in this context.

There seems to be no reason, therefore, not to take Weinberg's remarks at face value. A universe of fields, where "particles" (a Newtonian heritage) are simply creations of our minds, is something to wonder at and no less challenging to the imagination than the paradoxes and puzzles that we may generate through interpretations of quantum mechanics.

---

[21] Quoted from H.R. Pagels, *The Cosmic Code*, pp. 238–239, Bantam Books, New York, 1983.

[22] A sophisticated discussion of this issue, which concludes that "the position of a photon cannot be an observable," is to be found in Peierls' *Surprises in Theoretical Physics*, pp. 11–13, Princeton University Press, Princeton, NJ, 1979.

18

A Visit With Photons:
Identity and Flexibility

Because photons—quanta of light—are not subject to the Pauli exclusion principle, and because they have their source in Maxwell's equations of the electromagnetic field, they seem more fieldlike than particle-like; for electrons the reverse is true. As quanta of the electromagnetic field, photons share all the properties of that field. Because of the identity of quantum particles of a given species, they have no identity of their own; they are all indistinguishable from each other. Thus, each photon carries the seeds of all possible photons of the same energy. One cannot say "which" photon is in which energy state. This is what gives it the flexibility to pass through both of Feynman's double slits and interfere on the far side. Arguments based on the idea that two photons may be associated with different states are false.

Although in the previous chapter we insisted on the similarity between quanta of the electromagnetic field and of the electron field, there are important differences to be noted too. A similarity is that within each category quanta are indistinguishable from each other. A difference is that photons are bosons, that is, particles not subject to the Pauli exclusion principle, while electrons are fermions, which are governed by it. Bosons may be said to mediate the interactions between fermions, whereas fermions are seen as the "sources" of photon fields. This is a rather arbitrary way of describing the situation. Thus, electrons, through their electric charges, are a source of photons, which thus mediate the force between electrons. Similarly, nucleons (such as neutrons and protons) are a source of π-mesons, which in a similar way mediate the force between the nucleons. But the situation is more complicated than that, because there are positively and negatively charged π-

mesons, which can therefore also produce photons; in this way, one boson field can be linked to another.[23]

The simplest thing that we can say about the whole scheme of things is summed up in the previous chapter: that the physical world consists of the interplay of various quantized fields, and these are of two types, distinguished by whether or not they adhere to the Pauli principle.

Peierls[24] has, however, noted an important difference between boson (pho-

---

[23] At a deeper level, nucleons may be considered as bound systems of three quarks held together by gluons, while π-mesons are similarly bound pairs consisting of a quark and an antiquark. At this level, quarks are the fermions; gluons, the mediating bosons. Everything then is described in terms of quark–gluon interactions. The quarks have a new kind of change that makes them the sources of gluons. Again, there is a difference though: Whereas photons do not have a charge and so do not generate a new generation of photons, gluons themselves *do* have this new kind of charge and so are capable, by their very existence, of generating more gluons. The situation is in this case more akin to that of gravitational quanta (gravitons), which, because they carry energy (though not rest mass), can create more gravitational field (i.e., more gravitons).

In the case of both gluons and gravitons, the theory is said to be *nonlinear*; that is, the amount of boson production by a fermion is not proportional to the strength of the fermion charge. This is a source of formidable mathematical difficulties.

The situation is still more complicated in that still another kind of mediating particle (of the W-Z family) governs the "weak" interactions between light fermions (leptons, such as electrons, μ-mesons, neutrinos), and these new particles may have electric charge, thus making them sources of photons, as well as "quark charge" (otherwise known as "color charge"), which in turn link them to quarks and gluons.

At this point, the reader may feel as though drowning in a sea of detailed interactions and be inclined to ask whether the world is really that complicated." Unfortunately, the answer is positive, and one is left asking, "Where is the essential simplicity of which physicists like to boast?" A related question is, Is not all of this pretty arbitrary and "ad hoc"? Although there are links and unifying elements not evident at first sight, could not Einstein's God have done a more tidy job? Again, one is tempted to ask, Why not, indeed?

When the universe was born, back at the Big Bang, it seems that what came into existence was this sort of tangled maze. It makes one feel that those contemporary physicists whose avowed goal is to "explain" how all of this was created out of the simplicity of nothingness have somehow wandered into a world of pure euphoria, where they await some sort of divine revelation—a blinding light that melts all into an ephemeral simplicity.

But consider this; the "real world"—that which we experience—is, if not transparently simple, not nearly so complex as that which we have painted. The *necessary* components are relatively few: electrons, photons, nucleons, π-mesons, μ-mesons, and neutrinos. The rest is a substratum revealed by the limits of our technical capacity to create extensions to the world of our senses and the capacity of our imaginations to envisage an edge of reality hidden to our experience. There is still much mystery, and much wonder, closer to the realities that impinge on our daily lives.

[24] *Surprises in Theoretical Physics*, Sec. 1.3, pp. 10–14, Princeton University Press, Princeton, NJ, 1979. Peierls notes that there is a qualitative distinction between photons

ton) and fermion (electron) fields. In the photon case, one can start with the classical field and then quantize it; the quanta are particular *elements* or bite-sized samples of that field. For electrons, there is no classical field since, by the correspondence principle of Bohr, the classical and quantum cases merge only for large quantum numbers, whereas only single quanta can exist for electrons.

The consequence is that in the electromagnetic case the quanta are simply scaled-down samples of the classical field and share all of its other properties. This is an important consequence because there is no more restriction on the extent or geometry of the quantum than on its classical "parent." We must not think of photons as quantized plane waves, for example, with a definite momentum. In empty space, there can be monochromatic waves of many sorts: in the case of the Feynman two-slit experiment, waves that can go through two slits. There can be spherical waves, such as multipole waves radiated by an antenna, as well as plane waves. Pfleegor and Mandel[25] have shown that single photons may be created by combining waves of the same frequency from two different lasers! An essential feature of this phenomenon is that photons are completely indistinguishable. If you have a field, of whatever geometry, containing N photon's worth of energy, it does not consist of N identifiable photons. One photon's worth of energy can take place in an interaction, according to the constraints of the problem.

It was presumably this sort of consideration that led Dirac to his dictum: Photons cannot interfere with each other but only with themselves.[26] An example is to be found in Feynman's two-slit setup, where it is evident that *each photon* incorporates the interference pattern.

An image can be used to illustrate the point. Bartenders have a device which, when inserted into the cork of a whiskey bottle, allows them to pour out a definite quantity of whiskey. We may think of that quantity as a quantum of whiskey. Inside the bottle, all of the whiskey is identical; the quanta are not identifiable. Furthermore, the shape of the bottle (unless it were to have some peculiar topological features) does not matter. Though one cannot say that the quantum has the shape of the bottle, precisely because of its non-identifiability, it is, however, the quantum of a field of that geometry.

Another consequence of indistinguishability is to be found in a system of two free plane wave photons of the same frequency. No charge can interact with only one of these states, since their wave function must be a symmetric

---

and electrons: "There can be no *classical* field theory for electrons and no classical particle dynamics for photons." Thus, "wave-particle duality" does not take the same form in the two cases.

[25] R.L. Pfleegor and L. Mandel, Phys. Rev. **159**, 1084, 1967.

[26] P.A.M. Dirac, *Quantum Mechanics*, 4th ed., Ch. 1, p. 9, Clarendon Press, Oxford, 1958.

combination of the two; the interaction must therefore involve both. Is there a similar sort of phenomenon for electrons? In some circumstances, yes. A simple example is to be found in the states of the hydrogen atom. In p-states, for example, in the absence of a magnetic field, there is a three-fold degeneracy for states of each spin. Any linear combination of these three states is itself a valid state that can couple to a photon.

The important thing to note in both cases is that the linear combinations that we form must be made up of states of the same energy. *This is a criterion for quantum states in general: they must have a single energy or frequency; that is to say, everything must vibrate to the same frequency.* Only in this way can the coherence of the state be ensured. Such states, despite their apparent periodicity, are in effect stationary; the only physical change that can take place is a change from one state to another. Such changes, as already remarked in the case of electrons, involve charge densities and currents that couple to the photon fields and therefore become unstable. These comments are as true for many-particle as for one-particle states.

It follows that no information can be transferred within such a state; all information is in the quantum numbers of the state, which pertain to its global properties, which are collective and normally relate to internal symmetries or patterns. As with classical systems, which have normal modes of vibration, the pattern is maintained in the vibration. It is, however, a strictly *quantum* feature that the "vibration" does not imply physical change, since frequencies, being related to energy, are static properties. No internal physical processes are required to sustain the "vibration"; the system is static because the pattern is static.

19

Can Wave Packets Be Particles?

The wave function of a free particle is of infinite extent, which seems hard to reconcile with the notion of a point particle. One way to cope with this difficulty is to consider particles to be represented by "wave packets" with some degree of localization realized by superposing states of different frequencies in such a way that there is destructive interference outside a limited spatial range. Such "packets" cannot have a definite energy, and since the component states evolve independently, they are unstable and cannot maintain their form. From the viewpoint of quantum mechanics, each frequency can exist only in quanta of that frequency, which then interact independently. For purposes of interaction, then, the effect of localization is lost.

The fact that the wave function of a free particle of a specific energy must have infinite spatial extent and yet is spoken of as a particle has given rise to the idea of wave–particle duality, clearly a somewhat paradoxical notion. As noted, one resolution is to contend that the wave function does not describe the particle at all but only determines the probability that a point particle of given energy may be at any particular place. This is the probability interpretation of Born, which does not concern itself with the phase. Feynman's path integral approach, in which the phases of different paths interfere to explain physical phenomena, circumvents this problem but at the expense of denying *separate* paths any physical significance. Another more naive idea is to use wave packets, which are superpositions of pure states and which may be somewhat localized, as models for real particles. There are difficulties with this idea, however. One is that such packets, made up of components that evolve differently in time, tend to come apart as they evolve; they are not stable. This would hardly do for a particle like the proton, for example, whose lifetime is so long as to be, up to the present, unmeasurable. Another difficulty is that it does not have a well-defined energy, so that when it is involved in reactions energy can be maintained only if the energies of the other interacting components are equally imprecise, while maintaining precision in the total system. Thus, the nature of one particle would be dependent on that of the others.

Again, a way out of this puzzle is to evoke the probability interpretation, according to which the wave packet again merely describes the uncertainty of the *state* of the presumed underlying point particle. The implication is, then, that the particle interacts in one or other of the pure states out of which it is composed. Energy conservation can then be realized, but again at the expense of complete delocalization. One now has to evoke probability twice: the particle must first have a probability of being in a certain pure state and then in that state have a probability of interacting at a specific point. When proponents of this view are asked how this can be, the honest answer can only be that no one knows. But can one say in such a case that one really has an interpretation at all?

Unfortunately, the problem of interaction causes another difficulty: Interactions involve a transition from one quantum state to another. In the process of transition, the wave function must be in a state that involves a superposition of its initial state and its final state, and usually other transient states as well. Under such conditions, the various component states evolve differently in time, and this evolution depends on the relative phases of all of the components. Thus, an interpretation based only on probabilities is not possible.

Still, this view has one clear advantage: that it reveals that the "true" particle, such as it is (the particle involved in the primitive interaction with other matter), is in a pure quantum state. This fact is very important in the discussion in Chapter 21 of the so-called "delayed-choice experiment."

This suggests that it is legitimate to use the probability interpretation in interpreting impure or mixed states. Each of these states is itself a pure quantum state. But to assume that states exist in which position is a valid quantum number is to make a fundamental assumption about the very nature of matter, and one that is without experimental foundation. What is evident is that, in this instance, any valid interpretation must take account of the significance of phase relationships.

20

THE QUANTUM STELLAR INTERFEROMETER OF HANBURY BROWN

The conception of photons that we have expounded in Chapter 18 is illustrated in the stellar interferometer of R. Hanbury Brown. Photons of a given frequency emanating from the whole surface of a distant star and converging to the point of observation on Earth have a small lateral spread of momentum, depending on the size of the star and its distance from us. By the uncertainty principle, this gives the arriving photons a transverse dimension, which may be as small as a dinner plate or as large as many meters in diameter. Just as the convergence of the photons in Feynman's two-slit experiment give rise to an interference pattern at a detector, so do those from the star. Provided that two detectors are set up at a distance from each other less than the radius of the photons, they will show the interference pattern. In this way, the transverse dimensions of the photon can be determined, and from them, the diameter of the star can be deduced. The smaller (and more distant) the star, the greater the photon's lateral dimension will be.

The ideas of the nature of photons that we have developed are strikingly illustrated in an ingenious method adopted by Hanbury Brown to measure the diameters of stars too distant to be susceptible to direct measurement. The device that he adopted makes use of the quantum interference of photons of macroscopic dimensions. Two features are used: one is the use of the self-interference of macroscopic photons, just as in the case of the two-slit experiment. The other is the ability of a detector to register photons that are quanta of a macroscopic field created by multiple sources.

The source of the stellar radiation will be photons of a given spectral frequency from all of the surface of the distant star. Individual photons emitted at different points on the surface are not distinguishable but all meld into a single field. Out of that composite field, quanta of energy, photons, may be skimmed off by the detector.[27]

What is the character of the field arriving at Hanbury Brown's detector? It will converge from all points on the surface of the star facing us on Earth, that is, from a radiating area of πr^2 where r is the radius of the star. On this scale, Hanbury Brown's apparatus is, for all intents and purposes, a point target. The light from the star will then arrive with an angular (transverse) spread of r/D, where D is the distance to the star, supposedly known from other considerations. For a frequency f, the total momentum of the photons will be hf/c, so the transverse momentum will have a spread of $(hf/c)\,(r/D)$. We may now use the uncertainty relation between momentum and position to

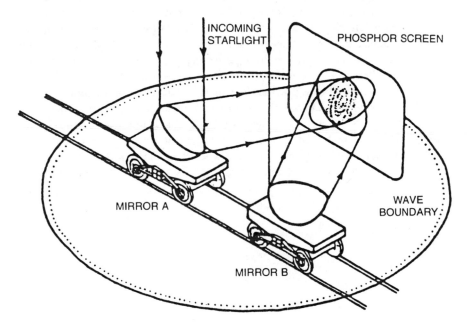

Figure 3.
Light from the distant star is reflected at mirrors A and B and is focused on a phosphor screen. There will be an interference pattern analogous to that in Feynman's two-slit experiment. Since the waves have a lateral spread depending on the distance and diameter of the star, there will be no interference if the spacing between the mirrors is greater than the width of the photon. Given the distance of the star, this permits a calculation of its diameter. (Adapted from N. Herbert, *Quantum Reality*, Anchor-Doubleday, 1985.)

----

[27] It is sometimes said that when someone in our age takes a breath of air, it will contain several molecules of Caesar's dying breath. Unfortunately, since all molecules of a given chemical substance are indistinguishable, the statement is really meaningless.

conclude that the photon will have an angular spread of $(hf/c)\,(r/D)$. This determines a transverse dimension of the photon of $hcD/2hfr = cD/2\pi fr$. For the star Betelgeuse, which is 520 light-years distant, and with light of 5000 Angstroms wavelength, it is a meter or so. For smaller stars, or ones at greater distance, the "size" of the photon will be correspondingly bigger. The apparatus, shown schematically in Figure 3, consists of two mirrors mounted so that the distance between them may be varied. The mirrors may be focused so that the images will coincide at a point on a sensitive screen, permitting them to interfere. If they are then gradually separated, the interference pattern will disappear, and no more photons will be registered, when the separation becomes bigger than the dimensions of the photon.

The situation is rather similar to that of the screen with the two slits. The images from the two mirrors correspond to those arriving from the two slits. It is unambiguously clear that the interference is that of single photons of macroscopic dimensions. As in the two-slit case, then, individual photons will be registered at seemingly random positions, the interference pattern only appearing when a large number have been registered.

The method may be used when the star is distant enough, or small enough, that its dimensions cannot be resolved by ordinary optical means. There is, of course, a limit. If the star is insufficiently luminous or too small or too distant, there may not be sufficient photons to create a distinct interference pattern.

The method owes its power and its validity to the quantum character of photons and so serves as confirmation of the ideas concerning photons that we have developed.

21

About the Delayed-Choice Experiment

In the so-called "delayed-choice" experiment, a beam of photons is split by a half-silvered mirror, so that parts of it can go by two different routes. After reflection, the two halves of the beam can be recombined by a second such mirror, and interference is observed. We are then asked to imagine that the second mirror is not inserted until the photons have already accomplished their travel, or after they have chosen one path or the other.

The fact that interference can still be observed is then interpreted not as an objective physical effect but rather as a creation of the experimental operation, which in turn shows that there is a conflict with the view that the universe exists "out there" independent of all acts of observation. The result is paradoxical in that it purports to show that our measurement determined what had happened in the past!

What assumptions have gone into this argument? First is the classical conception that the photons have such a character as to follow a trajectory, in which successive positions represent the position of the photons at successive times. Second is the assumption that the past history of the photon is created by our measurement, thus denying that there is an objective reality independent of our intervention.

We shall show that with other assumptions less devastating to the fundamental tenets of science a quite different interpretation can be found within the framework of quantum theory.

The delayed-choice "thought experiment" of John Wheeler has attracted much attention as a paradox involving serious difficulties for quantum mechanics. But once again, there is a paradox only if one adopts the conception of photons as point particles following a trajectory around an optical path.

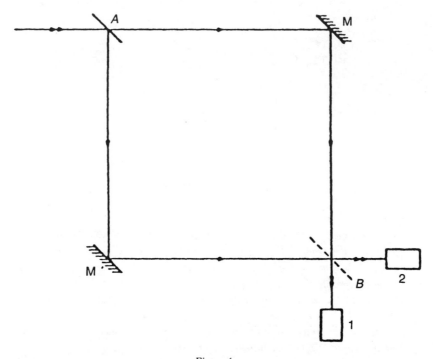

Figure 4.
The delayed-choice experiment. A weak beam of monochromatic light is split by a half-silvered mirror and reflected at mirrors M and M' to detectors 1 or 2, respectively. If a second half-silvered mirror is inserted at the crossover point B, the split beams recombine and create an interference pattern at detector 2. It is said that if the mirror B is inserted after the photons have passed the half-silvered mirror A, that is, after they are committed to one or the other branch, no interference should be observed. Since the photons actually go by *both* paths (just as in Feynman's two slits), and since they are *not* localized and so do not pass mirror A at a fixed time, interference still takes place.

Rather than posing questions about quantum mechanics, the fact that this interpretation leads to a paradox provides a cogent argument against the interpretation. As we shall see, the paradox evaporates if we argue strictly from quantum principles. What quantum mechanics, properly applied, predicts is precisely what is found.

The experimental setup is portrayed in schematic form in Figure 4. P.C. Davies, in *The Ghost in the Atom*,[28] describes the experiment thus:

An idealized experiment designed to carry out a [related] delayed-choice experiment...forms the basis of a practical experiment performed recently by Carol Alley and his colleagues at the University of Maryland. Laser light incident on a half-

---

[28] Cambridge University Press, New York, 1986, pp. 9–10.

silvered mirror divides into two beams so that they cross and enter photon detectors 1 and 2. In this arrangement a detection of a *given photon* by either 1 or 2 suffices to determine which of the two alternate routes the photon will have travelled.

If, now, a second half-silvered mirror B is inserted at the crossing point...the two beams are recombined, part along the route into 1 and part along the route into 2. This will cause wave interference effects, and the strengths of the beams going into 1 and 2 respectively will then depend on the relative phases of the two beams at the point of recombination....

Now the crucial point is that the decision of whether or not to insert the second half-silvered mirror B can be left until a *given photon* has almost arrived at the cross-over point. In other words, whether the photon *shall* have traversed the system either by one route or by "both routes" is determined only *after* the traverse has taken place. [emphases added]

The first point to note here is that, when the beam is divided, the parts of the wave on one limb are phase correlated with those on the other; otherwise interference would not take place. By the definition of *photon*, this means that they are parts of the *same* photon. This is surely independent of when the second mirror is introduced.

Photons that do not contribute to the interference effect may of course still be detected at either of the two detectors. The theory tells us that *only whole photons may be detected*. That this is possible requires only that there be an overlap between the photon wave and the electrons of the detecting apparatus. One may be tempted to say that these photons have gone by one or the other path. Such a statement ignores again the fact that photons are not distinguishable. As shown by the Pfleegor-Mandel experiment, the photon emitted by the laser source cannot be identified with one that is absorbed. All of the source photons are phase coherent; the interference experiment manifests itself when the experimental conditions (i.e., the second mirror) permit(s) it.

This is not to say that there is not something strange here. The strangeness is in the fact of the total absorption of the very nonlocalized photon when recombination does not take place.

It is important to note that, when interference does take place, it is not the case that two point particles have been recombined but rather two coherent waves with finite lateral extension perpendicular to their respective momenta.[29]

---

[29] Philosophical discussions often founder on the rocks of scientific detail. Even for the "idealized" experiment we are dealing with here, there are complications. For example, when a photon is reflected or transmitted by mirrors, the changes in phase that take place must be taken into account. Let us designate the parts of the split beam as "horizontal" and "vertical", "upper" and "lower", "left" and "right." One path will then be, when the original beam comes from the upper left, as follows: traverse upper left mirror; reflect at upper right; traverse at lower right to detector 1. Another path, which leads to detector 1 is as follows: reflect successively at the upper left, lower left, and lower right mirrors. These are not identical paths; one involves two

From footnote 29, we can see that, over a modest distance, a rather narrow beam has for all practical purposes a well-defined wave vector, as we know from experience with lasers.

However, photons in quantum states with well-defined momenta do not progress from point to point along a path; they do not have a "trajectory." The momentum of the wave is a characteristic of the wave as a whole; the same momentum density is present in all equal parts of the wave. We cannot therefore speak of the photon—the quantum of the wave—as moving along the path of the beam, in the sense of being first "here" and later "there." So statements like those made by Davies and Wheeler—"whether or not to insert the second half-silvered mirror can be left until a given photon has almost arrived at the cross-over point" (Davies op. cit.) or "the new feature about the delayed choice version of this experiment is that we can wait until the light or photon has accomplished almost all of its travel before we actually choose between the photon going by both routes or a photon going by only one of those routes" (Wheeler op. cit. p. 65)—make no sense.

What Wheeler and Davies are in effect saying is that quantum mechanics is guilty of a violation of causal relationships; what happens later can affect what happened earlier. But this violation lies not in the structure of quantum mechanics but with an untenable interpretation of it.

It all comes down to the fact that, in a pure quantum state, no information is carried internally from one point to another. We know, of course, that waves can carry information, but only if they combine a range of frequencies. Consider a musical note at a sharp individual frequency; we learn nothing from it. Nothing changes with time. It is constant, monotonous. Information carried by sound (voice, music, etc.) gives us information in a temporal sequence. This is brought about by the superposition of many frequencies, whose phases change at different rates. In quantum mechanics, states are characterized by definite frequencies. In the case of light, a superposition of frequencies means a superposition of photons and their energy; one photon carries only one frequency and a fixed quantum of energy.

---

traversals and one reflection, and the other three reflections. These will not recombine in the same way as the limbs that recombine to reach detector 2. One of these latter paths is as follows: traverse upper left mirror, reflect at upper right, reflect at lower right, while the other takes the following path: reflect at upper right, reflect at lower left, transmit at lower right. These two paths should recombine in phase, both involving two reflections and one transmission. If the two paths were of the same length, there would be constructive interference on recombination.

Then there is the factor of lateral spreading. This involves considerations like those in the Hanbury Brown and Twiss device. If the beams are confined to a lateral width of w, this will involve an angular spread of $c/wf = L/w$, the frequency being f and the wavelength L. The lateral spread of the beam in a distance d as a fraction of its original width is Ld/w^2. For a well-defined beam, the beam width must be $\gg (Ld)^{1/2}$. One can only propagate over substantial distances beams that are not too narrow. On the other hand, as beam width increases, beam intensity decreases.

Electrical engineers can create electrical signals of all sorts of shapes. They do this by mixing many frequencies; that is, by creating a pattern of photons. They can make something temporally localized, a "click," by superposing a wide range of frequencies. All that the listening device "hears" is the net electric field of this mix of various frequencies. In the case of sound, the human ear and nervous system somehow untangle this mix and detect well-defined frequencies in it. Similar phenomena exist for light and the human eye. What strikes the retina of the eye are photons of various frequencies. The eye must unscramble them; it is the unscrambling of the effect of many photons, each of which has a distinct frequency or "color." Thus, an electrical signal can be propagated from one point to another, but to do so, it must consist of a spectrum of frequencies that cause the electric field to vary in a complex way. A single photon carries only a single frequency, so that when we reduce the intensity of the signal to the point that we can detect individual photons, we lose the whole pattern of information in the signal. Each photon is only an endless buzz. The electric field does not change, nor does its pattern go from place to place.

The delayed-choice experiments are designed to detect only individual photons, unchanging and monotonous. By their nature, then, they are as likely to be detected (i.e., absorbed or changed by some agent) at one time as another. The moment at which this happens is a matter of pure chance. It does not mean that their detection tells you where they are; only that they have changed into something that they were not before. The randomness is in their interactions, which is all that can be detected about them.

The delayed-choice experiments therefore do not involve a paradox. Whenever a photon is detected, it carries all the information that is inherently present in its structure and must therefore, by virtue of the nature of this experiment, show the interference pattern. That is, of course, exactly what the experiments show.

22

The Illusion of Superluminal Signaling

In two other paradoxes, the probability interpretation of the wave function appears to produce behavior contrary to known physical principles. The first appears to suggest superluminal signaling, that is, the transmission of signals at speeds greater than that of light. The other is the principle of conservation of fermions (in this case electrons). Davies' split-box gedankenexperiment involves what is usually known as the "instantaneous collapse of the wave function," in violation of the principle of relativity. However, the speed of light is not the central issue in this case, but rather the fact that the conventional interpretation involves the concept of the instantaneousness of distant events, which is meaningless in relativity. Happily, quantum mechanics itself removes the paradox, since in the presence of an asymmetry of the system, however minute, it ensures that not only the electron but its wave function cannot be subdivided.

The other paradox, which involves the modified EPR (Einstein-Podolsky-Rosen) experiment suggested by Bohm, has to do with the effects of the quantum "entangling" of two particles, in this case photons. Here again, it appears that a measurement at one point can affect the results of an observation at another distant one. But whereas the usual interpretation is that the effects concern two separated systems (photons), the identity of photons requires that the combined (or entangled) state of the two behaves as a single system, in which the photons detected by measurement are found with equal probability at either site.

22.1. DAVIES' SPLIT BOX AND WAVE FUNCTION COLLAPSE

P.C.W. Davies has edited a book[30] designed to inform the general reader of "the mysteries of quantum physics." In it, Davies discusses a number of the traditional paradoxes that are said to plague quantum theory.

One of these puzzles has to do with the supposed "collapse of the wave function" and the fact that it opens the possibility of instantaneous communication at large distances, in contradiction to the requirements of the theory of relativity.

The illustration that he uses is a very artificial and unrealistic one. He asks us to envision a box, within which is imprisoned one electron. The nature of this box is not specified, though it surely must not be intended to have electrons as constituents, because one could not then distinguish the trapped electron from those constituting the box—a complication, surely. We shall for the moment overlook this little difficulty; we shall suggest later a more realistic system exemplifying the same principles.

Let the box be rectilinear. The electron wave must fulfill the condition that there must be an integral number of half-wavelengths in each dimension of the box. The lowest energy state is that in which a single half-wave fits into each of the box's three dimensions.

Suppose now, with Davies, that an impenetrable barrier is inserted to cut the box into two separate identical compartments. By symmetry, one half of the original wave function should be left in each of these compartments. The behavior of the wave should not be altered in the two transverse dimensions, but in the third there should now be two waves with half the wavelength of the original (Figure 5). A reduction in the wavelength by a factor of 2 implies an increase in the electron energy in that direction of a factor of 4, so this exercise implies a substantial input of energy. This energy would then have to be expended to insert the barrier. The main problem is, however, that it implies the possibility of cutting one electron into two of higher total energy. This violates a basic principle of quantum mechanics: the principle of conservation of leptons, a family of particles of which the electron is a member. There appears to be a way around this difficulty, however; it lies in Born's probability interpretation, according to which the wave function is not itself a physical entity but only determines the *probability* that the (point) electron will be found at a given position. We cannot say that the electron *is* in one compartment or the other; only that it is equally likely to be in one as in the other. Please note that we are dealing here with contingent probabilities; it is not permitted that the electron be in neither box or in both. In other words, the two parts are still taken to be separate parts of a single whole, just as was the case with the double-slit setup discussed earlier. If, therefore, you were

---

[30] P.C.W. Davies and J.R. Brown, *The Ghost in the Atom.*

bothered by the idea that a single particle wave function could comprise parts apparently disconnected from each other, this is also involved in Davies "conventional" treatment of our present problem!

We now imagine that the two compartments are removed from each other, perhaps as far as a light-year. Yet the two parts are still viewed as being intimately connected with each other. Imagine then that one of the boxes is now opened and the electron is found in it. Then according to the conventional Copenhagen interpretation, this observation, or measurement process, has *put* the electron in that state. It then follows, says Davies, that "Instantaneously, the quantum wave in B [the other compartment-author] vanishes, even though it is a light year away." This is an example of the process known as collapse of the wave function. The logic seems cogent; there must, from that moment, be no possibility of finding the electron in B, by virtue of conservation of leptons.

Yet there is a glaring flaw in the argument, for relativity tells us that there is no such thing as simultaneity at remote points; it is a concept without meaning. If two events are simultaneous in one frame of reference, in a frame of reference moving uniformly relative to the first, one event may precede or follow the other. To put it differently, suppose that the electron is found in compartment A. This cannot affect the situation at the compartment B until a signal from A reaches it, and this will take a light-year. In the meantime, since the wave function at A has not yet vanished, there remains a possibility of one half that an electron will be found at B if that compartment is opened. Thus, two electrons will have been created from one, and conservation of leptons will be violated.

This is not a new argument. Feynman uses it[31] to demonstrate that conservation laws must always be local; that is, the conservation must take place locally, at each point.

What conclusion does one draw from such a demonstration that the notion of the collapse of the wave function is not consistent with the requirements of relativity? Some will respond, Yes, that is a problem; quantum mechanics implies instant communication at large distances, and so is inconsistent with relativity. Since both relativity and quantum mechanics are incontrovertible theories, physics is in a state of deep crisis. But surely this is a radical and irrational conclusion, for the structure of quantum mechanics does not contain the hypothesis that wave functions must be interpreted as *probability amplitudes*, nor is *collapse of the wave function* mandated by it. Is not the natural and logical course to reject the interpretation that made quantum mechanics seem paradoxical?

A more natural analysis is possible. We do not really believe, in our bones, that an electron can be cut in two. We are forced, in fact, to acknowledge that

---

[31] *The Character of Physical Law*, p. 64.

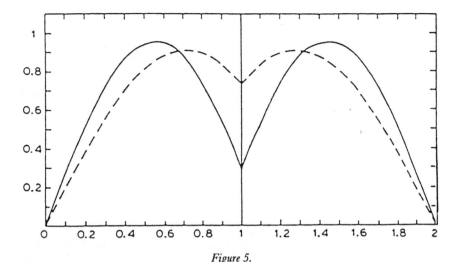

Figure 5.
The wave function when the barrier is inserted at the precise midpoint; the dotted curve is for a weak barrier, the solid one for a much stronger one. For any asymmetry, however minute, the symmetry will be broken and the symmetric state portrayed will become degenerate with the first excited (antisymmetric) state to concentrate the wave function on one side or the other of the barrier, as shown in Figure 6.

the electron really is in one or the other box long before the vast separation is carried out. This can in fact be realized within the framework of quantum mechanics itself, provided that we admit the fact that *perfect* symmetry is not realizable in nature.

The situation is the following:[32] Let us assume that we have a barrier that is somewhat penetrable situated at *exactly* the midpoint of the box. Suppose now that the barrier were to become more and more opaque. The wave function would then evolve as shown in Figure 5, becoming in the limit of total opacity two half-waves of half the original wavelength. The first excited state in the box would be a complete sine wave, the same as before, except that the wave function would be reversed in sign on the right-hand side. Thus, with a totally opaque barrier, the two states would have the same energy (i.e., would become degenerate). If the barrier is displaced by a minute amount, creating a very weak asymmetry, the wave function evolves as in Figure 6 until, in the limit of total opacity, the whole wave function is found on one side of the barrier.

A similar result is found if the barrier remains at dead center and any small perturbation on one side or the other breaks the symmetry. In the limit,

---

[32] The mathematical calculation is sketched in my paper "Is Quantum Mechanics Paradoxical?" Phys. Essays **4**, 614–618, 1990.

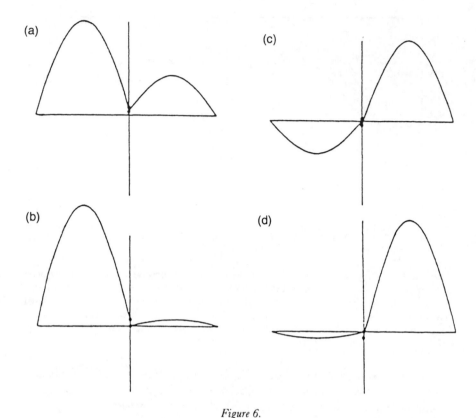

Figure 6.
(a) and (c) The wave function for a weak barrier. (b) and (d) The wave function for a strong barrier. For a totally impenetrable barrier, the electron will be concentrated on one side or the other.

the asymmetry breaks the degeneracy between the even and odd states, resulting in states that are even and odd linear combinations of the two states. Again, the resulting states show the particle totally on one or the other side of the barrier. Thus, it appears that in the real world quantum mechanics ensures that the electron will truly find itself located in one or the other compartment by the time the barrier becomes impenetrable.

There is an equivalent situation in molecular physics. The hydrogen molecule ion has one electron circling around two hydrogen nuclei. This could be formed by the junction of a hydrogen atom and a hydrogen positive ion (i.e., a bare proton). Such an ion is stable when the nuclear separation is about an Angstrom and the binding energy is about one fifth that of the separate hydrogen atom. In this state, the wave function is symmetric about the midpoint between the two nuclei. We might ask whether there is a state that is antisymmetric about the midpoint of the nuclei. Calculation shows that such a state will not be bound; its energy will decrease as the nuclei separate. In fact, as they separate, it will soon be found that the energy will be lower if the electron stays by one or the other nucleus, becoming at large separation five times lower than in the molecular ion! It is clear then that the symmetry of the ion will be *spontaneously* broken as the nuclei separate, since the asymmetric state will tend at large distances to become degenerate with the symmetric one, and once again the lowest energy states will be those localized around a single proton.

In this more realistic problem, it seems intuitively obvious that, if we separate the two protons, the result will ultimately, at sufficiently large separation, be a hydrogen atom and a proton. That this involves a spontaneous breaking of symmetry is an afterthought, not a dominant factor in the problem. We may ask, with slight surprise how that happened. But such spontaneous symmetry breaking is common in modern physics, and we would hardly give it a second thought. Since it is clear that symmetry is broken in the formation of the ion, why should it not be in the reverse process of separation? All we need believe in is time reversibility of fundamental processes, which is normally taken for granted.

22.2. The Modified EPR Experiment With Photons

The misgivings of Einstein about quantum mechanics are well known but perhaps not so well understood. A point that remains unclear is the extent to which these reservations spring from the standard "probability interpretation" of the Copenhagen school and Einstein's acceptance of this interpretation as inseparable from the theory itself. Einstein was concerned about the *reality* of quantum concepts, like, for example, the position or the momentum of an electron. Despite the uncertainty principle, which stated that these quantities were not simultaneously determinable, he felt that, as elements of reality *for a particle* they must separately exist, so that the uncertainty principle itself should have some explanation in realistic terms; the fact that it did not seem to exist appeared to him to be an indication of the incompleteness of the theory. In a paper in *Physical Review* in 1935, Einstein, Boris Podolsky, and Nathan Rosen contended that if these attributes had a reality, they could, in contradiction to the uncertainty principle, both be measured.

In the same sense that it is true of the Davies split-box experiment that it is based on an idealized and unrealistic situation—which, however, reveals the basic issues—the *gedankenexperiment* envisaged by EPR, which postulates two distinguishable electrons separated by a large distance and then subjected to separate measurements of their position and momentum, contains too many incongruities to make a convincing argument. David Bohm suggested a more realistic experiment, which raised precisely the same issues.

The experiment concerns two photons emitted in the mutual annihilation

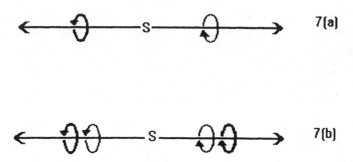

Figure 7.

Two plane-wave photons emitted in opposite directions from electron-positron annihilation. (a) The first interpretation, which leads to paradox, has one photon going in one direction and the second in the opposite direction. However, it is not the photons but the states that have opposite directions. Due to the indistinguishability of photons, either of the photons may be in either state, as portrayed in (b).

of an electron and a positron in a positronium "atom" in its lowest energy state. In the rest frame of the system, two photons will be emitted in opposite directions; furthermore, since the angular momentum of the system is zero, these photons will spin in the same direction about their respective directions of motion; that is, they will have the same circular polarizations. The argument then states: If one measures one of the photons to have right circular polarization (say), the other will be found also to have right circular polarization. If one measures the *linear* polarization of the first, the second will be found to have the same linear polarization. But a linearly polarized photon is a linear combination of states circularly polarized in opposite directions. Thus, the measurement made on photon 1 will affect that subsequently made on photon 2, even though they may separate to an arbitrarily large distance. Thus, again, relativity seems to be violated.

Sakurai[33] has treated this problem in detail; we shall simply describe some aspects of it from his discussion. It is apt first to call attention to a remark of G. Källen, which Sakurai quotes, "We should like to warn the reader against too pictorial an interpretation...in terms of particles propagating from one space-time point to another since such pictures sometimes lead to serious misunderstanding of the physics involved." But we go on to note the manner in which Sakurai describes the system of emitted photons. Operators like $a^{\dagger}_{k,1}$ represent the creation of a photon, which we label *photon 1*, which has a momentum $\mathbf{k}$. Similarly $a^{\pm}_{k,2}$ represents a second photon (photon 2) with opposite momentum $-\mathbf{k}$ (see Fig. 7a). Because they are identical, each pho-

[33] See *Advanced Quantum Mechanics*, pp. 211, 212, Addison-Wesley, Reading, MA, 1967.

ton is as likely to be in one momentum state as the other. The two-photon system created by the annihilation of electron and positron is then defined by the two-particle state

$$(a^+_{k,1}a^\pm_{k,2} + a^+_{k,2}a^\pm_{k,1})|0\rangle,$$

where $|0\rangle$ is the photon vacuum state (Fig. 7b). That two operators are combined in this way is due to the fact that the photons are indistinguishable, so that we cannot identify a single photon with either of the two one-particle states. This is sometimes referred to as the *entanglement* of the photons. The two-particle state has the startling property that each photon has equal probability of going to the right or the left. *This distinction between photon and state is of critical importance. It is the photon and not the state that is detected (i.e., interacts with the detector) in the measurements involved in the experiment.* This is what the mathematics tells us. It is essential to emphasize this difference. Once again, the fact of indistinguishability has consequences that we may find difficult to grasp, as intimated by Källen.

What this means, of course, is that the two detected photons are not remote from each other; they both share the same states. Thus, the seeming contradiction with relativity is avoided.

Actually, the situation is even stranger than that. Sakurai *defined* the problem as that of two photons with opposite momenta; that is, going in opposite directions. The truth is, however, that the direction in which they are emitted is entirely indeterminate. The annihilating electron–positron pair constitute a sort of antenna, which is capable of radiating in any direction; in fact, if enough such events are detected (for instance, by using a surrounding detector, or rotating detectors in arbitrary directions), the radiation will be found to be isotropic. The real problem then ought to be that of *spherical* waves. Their only difference will be that they will carry opposite angular momenta. But now we see that the wave functions of the two photons actually overlap everywhere!

What a difference a change of mental images makes! The preceding argument is made by thinking of the properties of physical waves. If we try to recast it in terms of probabilities, it does not appear to be the same argument and seems much less convincing, especially if we do so with our minds attuned to the notion that we are dealing with particles following a trajectory.

23

THE MEANING OF BELL'S THEOREM

Einstein's insistence that, in any "reasonable" theory, whatever can be measured must have physical reality led him to the conclusion not that quantum mechanics was wrong but that it was incomplete. This led to the idea of "hidden variable" theories, that is, theories that satisfied his criteria and could be used to "explain" quantum phenomena. John Bell set out to derive specific consequences of any such theory. He derived a theorem—an inequality—that would have to be valid for it. Experiments undertaken to test this prediction showed that quantum mechanics violated the inequality. Aside from showing that quantum theory was not consistent with Einstein's criteria, it appeared to show that hidden variable theories could not be valid.

Reaction to Bell's finding was strong. Some, the philosophically minded, believed that it represented an important advance in physics; for others, pragmatically minded working physicists, it was simply a confirmation of quantum theory. In any case, it generated much debate and controversy. Was quantum theory really "queer," or did it only seem so because we had not fully understood the implications of quantum principles, or, as Max Born had suggested, had we inadvertently made some unwarranted assumption which had led us astray?

John Bell, a theoretical physicist at CERN interested in the philosophical basis of quantum theory, undertook to find the distinguishing feature differentiating it from classical Newtonian theory. This clearly bears upon Einstein's reservation about quantum mechanics and his long-standing polemic with Bohr. The distinguishing points on which he focused were *realism* and *localization*. We must adopt again a critical attitude toward the meanings attributed to these terms, since failure to do so may lead to mistaken conclusions about the realities of quantum theory itself.

The meaning of *realism* that we shall adopt is that enunciated by Einstein.[34] Where does quantum theory stand according to this criterion? In the first place, it does not permit two complementary variables to be "simultaneously measured or predicted"; this is an ontological statement. Second, because of the identity of particles, it does not permit the distinction between two systems referred to in the EPR paper (see Chapter 22), but rather treats them fundamentally as indistinguishable parts of a single system. To put it differently, the sort of two-particle wave function incorporating identity $\Phi_a(1)\Phi_b(2)$ $[+/-]$ $\Phi_a(2)\Phi_b(1)$ denies an independent existence to the state of one or other particle. The question is whether the existence of identity of particles of a given sort makes them any less "real." The point here is that in truth Einstein's realism has nothing to do with the state of reality as generally understood at all, but only defines what EPR are willing to *accept* as real according to their view of the world.

However, the important thing to note is that quantum mechanics does not meet this criterion for the validity of Bell's theorem.

The second of Bell's hypotheses is *locality*. Here again, things are not as simple as they at first appear. In the first place, *particles* in quantum mechanics are not *localized*. However, the definition of a local theory does not relate to that. Polkinghorne[35] defines locality in these terms: If two systems have been for a period of time in dynamical isolation from each other, then a measurement on the first system can produce no real change in the second. This raises a question, What do we mean by "two systems"? How can we be sure that what we regard as *two* systems are not two parts of *one* system? We have already noted (Chapter 10) that Bell regards the word *system* as having no place in the formulation of the foundations of quantum mechanics. This is the more so when the word is used without consideration of what a system is.

In the case of the two-slit problem, the particle responsible for the diffraction pattern is macroscopically nonlocal, but then, so is an electromagnetic wave. If there is novelty here, it is that in quantum theory we use the word *particle* for this very nonlocalized entity. What goes through the slits is not two independent entities but one, characterized by macroscopic phase coherence. Interactions can only take place with it as a whole. This is certainly a form of

---

[34] Abraham Pais, in his scientific biography of Einstein entitled *Subtle Is the Lord* (pp. 455–456, Oxford University Press, New York, 1982), quotes from the aforementioned paper of Einstein, Podolsky and Rosen: "We could not arrive at our conclusion if one insisted that two ... physical quantities can be regarded as simultaneous elements of reality *only when they can be simultaneously measured or predicted*. ... This [simultaneous predictability] makes the reality of [two complementary variables] depend upon the process of measurement carried out on the first system which does not affect the second system in any way. *No reasonable definition of reality could be expected to permit this.*"

[35] J.C. Polkinghorne, *The Quantum World*, p. 73, Princeton University Press, Princeton, NJ, 1984.

nonlocality, but its context is limited to single "systems" defined only in their entirety.

How about the split box or the hydrogen molecular ion? Here, the question is not a philosophical one but a practical one. Can quantum coherence be maintained if a system is subjected to radical external intervention? The answer is qualified—perhaps, sometimes, within limits. There may come a point when it costs nothing, energetically, to break it, as in the split box. On the other hand, in the case in which a beam is split by mirrors and rejoined, coherence can be maintained over macroscopic distances, as it is in the two-slit case. In the modified EPR experiment, it may be maintained in a two-particle system, again over macroscopic but limited circumstances; however, identity of particles again prohibits two particles from being distinguished as two systems.

We can therefore give only a qualified answer to the question of whether quantum theory respects the locality principle. The answers to some of the basic questions seem to be neither "yes" nor "no" but "do not apply" or "irrelevant."

The moral here is that we cannot use the criteria of classical physics to judge either the reality or the validity of quantum phenomena. It is basically this point that Bell's theorem serves to establish. The quantum world has its own realities, both more sustainable and more universal than those of the classical world.

That said, what is the essence of the Bell theorem (or *inequality*)?

Bell has demonstrated that, without specifying the details of any special theory, certain limits can be put on the correlations between different measurements of the polarization of the photons emitted in the modified EPR experiment referred to earlier. In the experiment, the correlation is first determined between measurements made for direction **a** for one of the photons and **b** for the other. Then the direction of the first is changed to **a'**, and the correlation measured again. Finally, the second is changed to **b'**.

For each orientation of the polarizers, four possibilities exist: A measurement along **a** yields the result $+1$ if the polarization is found parallel to **a** and -1 if it is found perpendicular. We designate as $P_{\alpha\beta}(\mathbf{a}, \mathbf{b})$ the probabilities of the different results. Each of α and β may be $+$ or $-$ according to whether the polarization is found to be $+1$ or -1. The correlation coefficient of the measurements is then

$$E(\mathbf{a}, \mathbf{b}) = P_{++}(\mathbf{a}, \mathbf{b}) + P_{--}(\mathbf{a}, \mathbf{b}) - P_{+-}(\mathbf{a}, \mathbf{b}) - P_{-+}(\mathbf{a}, \mathbf{b})$$

(positive correlations are $+$, and negative correlations are $-$; this explains the signs of the terms on the right). Bell's theorem says that the quantity

$$S = E(\mathbf{a}, \mathbf{b}) - E(\mathbf{a}, \mathbf{b}') + E(\mathbf{a}', \mathbf{b}) + E(\mathbf{a}', \mathbf{b}')$$

must lie between 2 and -2.

Let us be quite clear about the assumptions that have gone into this theorem. It will be true, regardless of theoretical details,[36] provided only that the hypotheses of Einsteinian realism and localization are accepted.

If the same correlations are calculated from quantum theory, each separate correlation will be $-\cos\phi$, where ϕ is the angle between the two directions of polarization measured by the two detectors.

This reflects vividly the fact that quantum mechanics does not satisfy the "local realistic" criteria.[37] A local realistic theory would not allow any correlation between measurements made by the two detectors, in the sense that these measurements would be completely independent of each other; only in this way would each be, in the curious terms of this discussion, "real." The joint probability of the two measurements would, in this case, be simply the product of the independent probabilities of obtaining the results of the two separate measurements. (It is a curious deduction indeed, made in footnote 37, that human consciousness can prevent nature from producing identical particles!)

In the quantum case, the angle between the two directions is the difference of the angles that they make with a fixed direction: Let us call these angles δ^1 and δ^2. Now from simple trigonometry,

$$\cos\phi = \cos(\delta^1 - \delta^2) = \cos\delta^1\cos\delta^2 + \sin\delta^1\sin\delta^2,$$

which is *not* simply the product of two terms, each referring only to one or the other of the detectors.

The critical experiments, made to resolve the conflict between nonexistent "locally realistic" theories and quantum mechanics, were made by Alain Aspect and his collaborators.[38] They found that the Bell inequality was strongly violated and that, when the angles between the four directions a, b, a' and b', respectively, were taken to be 22.5 degrees ($\frac{1}{4}$ of 90 degrees), the quantity S defined above was found to be 2.697, well outside the limits prescribed for local realistic theories—and precisely the value predicted by the quantum calculation. Thus, the entanglement of states so critical to quantum mechanics was verified.

---

[36] This is fortunate, because no specific theories are known satisfying these conditions!

[37] But this does *not* justify the subtitle of an article by B. d'Espagnat in *Scientific American*, November 1979: "The doctrine that the world is made up of objects whose existence is independent of human consciousness turns out to be in conflict with quantum mechanics and with facts established by experiment." This statement is the more muddled and incomprehensible in the light of the fact that the experiments referred to confirm beyond doubt the validity of quantum mechanics and *disagree* with what would be predicted by any "local realistic" theory!

[38] A. Aspect, P. Grangier, and G. Roger, Phys. Rev. Lett. **49**, 91, 1982.

24

Symmetry and Point Particles

The whole of modern physics is permeated by the conception of symmetry. It should be noted that the word has, for physicists, a quite precise meaning expressible in mathematical terms and not necessarily conjuring up the same vision as it does among laypersons. It has to do with the fact that a system does not change (i.e., is invariant) under certain operations—such as rotations, translations, reflections, and exchanges—but also other more esoteric ones. It may be recalled that Noether's theorem associated a physical conservation law with each symmetry operation. Thus, the quantum numbers of quantum states of matter reflect symmetries, and the symmetries of the wave function in turn have specific characteristics. This is clearly true of the chemist's molecules, and the solid-state physicist's crystals, both structures dictated by quantum laws. It is less transparently true of the classification of fundamental particles into families by their symmetry characteristics.

The symmetries of wave functions are not fully expressed in probability distributions; certain states (so-called p-states in atoms) are antisymmetric, yet the quantum probability distributions are symmetric. This reflects the fact that the phase is an integral part of the symmetry pattern. What is it that determines the symmetry of a probability distribution? Classical physics has an answer to that question, but the probability interpretation of quantum mechanics has not. If the wave function does not have a physical reality, the answer cannot be found in the wave function but should be explained by a process from which the wave function springs (hidden variables again?). But the process should at the same time explain phase relationships. The questions can be answered mathematically, but is not the function of an interpretation to explain in words what the mathematics expresses in symbols?

The symmetry of molecules, like that of the states of atoms themselves, is a fundamental characteristic of their quantized states. In fact, the quantum numbers, which give a complete description of the states, are all related to symmetries. The quantum numbers specify the values of certain dynamical variables whose conservation is linked, by a theorem that is due to the mathematician Emmy Noether, to specific symmetries. This provides one of the most important links between classical and quantum physics.

Noether's theorem is a very general and fundamental one. What it says, essentially, is this: If a physical system is invariant under some symmetry operation, it obeys a corresponding physical conservation law.

Suppose, for example, that the system is invariant under time translation. This means simply that if it behaves in a certain way today, the same system in the same circumstances will do the same at any future time. The invariance will be manifested in Schrödinger's equation. If the equation does not change when the time is incremented or decremented, time translation invariance is established. The corresponding physical law that applies in this case is that the energy in the system will be conserved.

If the system is invariant under translation in space, i.e., is taken to a different location, the statement is that its momentum will be conserved. Similarly, if it is invariant under rotation about a fixed axis, its angular momentum about that axis will be conserved.

There are more complicated and less familiar examples in particle physics, for example, where more exotic symmetries exist.

Noether's theorem was originally proven for classical Newtonian mechanics. However, it reappears in quantum mechanics in a somewhat modified form. Here, symmetries are related to quantum numbers, the set of numbers completely describing quantum states. The mathematical discipline dealing with symmetries is called *group theory*, which therefore takes on a particular importance in the study of quantum phenomena. The study of the symmetries of systems then provides the clues to their quantum description.[39]

Take, for example, the angular momentum about the nucleus of an electron in a hydrogen atom. The Hamiltonian operator (energy operator) is symmetric, that is to say, is invariant, under any rotation about the nucleus. This implies conservation of angular momentum, which is manifested also in the quantum state and expressed in terms of the quantum number(s) specifying that dynamical variable. The states do not necessarily have as high a symme-

---

[39] It is interesting to relate this to the fact that two components of the angular momentum do not commute, so that they cannot be simultaneously specified. One component of the angular momentum can be a quantum number if there is rotational symmetry about that axis. But the only case in which there can be symmetry about two different axes at once is if the system is symmetrical about all axes, that is, has spherical symmetry, in which case they can be simultaneously specified because they are both zero.

try as the Hamiltonian and thus the Schrödinger equation, yet they must still have the symmetry springing from the conservation law. The Schrödinger equation has spherical symmetry in the hydrogen atom, but its quantum states may have either symmetry or antisymmetry about the nucleus. This is referred to as a *spontaneously broken symmetry*.

Let us look more closely at the states of the electron in the hydrogen atom. The Schrödinger equation is spherically symmetrical about the nucleus, which exerts the same force (at a given distance) in all directions. From this rotation invariance, the Noether theorem tells us, conservation of angular momentum follows.[40] This is true in both classical and quantum theories. But while in classical physics the electron will circulate in orbits, which will not have the symmetry of the atom, in the quantum case there will be wave functions whose quantum designations will have discrete values. The wave functions themselves will have specific symmetries but will not be invariant under rotation. On the other hand, physically identical states of lower symmetry may be obtained from each other by discrete rotations. All that changes is the point of view; the new states have the same shape as the original ones but viewed from a different angle.

The reason for the discrete quantization of angular momentum is quite similar to the reason for normal mode frequencies on a violin string. Because the string is fixed at the two ends, it will only sustain waves having nodes at the two ends. In the rotational case, the system is unchanged by a rotation of 360 degrees or any multiple thereof; thus, the span of the wave function is 360 degrees; after such a rotation the wave function returns to its original value. This implies that there must be an integral number of wavelengths in that angular range.

If we pick some direction of rotation, symmetry will ensure that the angular momentum about this direction will have quantized values, which in fact turn out to be integral multiples of $\hbar$. This is analogous to the classical situation in that the *states* of the atom have lower symmetry than the equation that generated them. Whereas Schrödinger's equation is spherically symmetrical, and hence invariant under rotation around *any* direction, the states are only invariant around a *particular* (though arbitrarily chosen) direction. One of the unexpected features, however, is that the lowest quantum state is that in which there is no angular momentum at all, whereas classical orbits must always have finite, though unquantized, angular momentum. The lowest energy, or ground, states involve only radial waves and thus are rotationally invariant. Otherwise, the components of angular momentum about a particular axial direction are constant in the quantum states and are thus valid

---

[40] The classical angular momentum is proportional to the mass of the electron and to the rate at which it sweeps out area in its orbit, and is thus constant, as first noted by Johannes Kepler.

quantum numbers characterizing those states. The energy of the electron is also a valid quantum number. It is easily verified that the energy operator and the operator for any component of angular momentum commute, which is a necessary condition for them to be compatible dynamical variables for the state.

If, however, we determine the commutator of the angular momentum operators about two *different* directions, we find that it is not zero. Thus, two different angular momentum components cannot be simultaneously specified; there is an "uncertainty relation" between them. In fact, if the angular momentum in one direction is given, that about an axis perpendicular to it is totally indeterminate. A consequence is that the total angular momentum cannot be determined by adding vectorially its values about the three perpendicular (axial) directions. It can have no fixed direction in space.

The example of quantum chemistry appears to give physical substance to the wave function carrying the symmetry. What is difficult to understand, however, is how this fact can be reconciled with the interpretation of the wave function as simply providing a probability distribution for point particles. For example, so-called p states are spatially antisymmetric with respect to a definite axis, though they generate a spatially symmetric probability distribution. An asymmetric perturbation creates a linear combination of a symmetrical s state and an antisymmetrical p state; this produces a spatially biased probability distribution weighted to one side of the nucleus, which is essential for molecule formation. This process cannot be described in terms of probabilities alone, for in each individual state, whether of s or p character, the probability distribution is symmetrical. The electron therefore carries not only probability information but also relative phase information, which plays an essential role in determining where it may be. Thus, the wave function seems to have a physical role and not merely an epistemological one. The amplitude of the wave function is interpreted in terms of the probability distributions of point particles, but interference depends on phase variations of waves. Yet in chemistry the probability distribution also depends on the phases of wave functions. No explanation can be given for the interference of probability distributions or for the fact that a physical phenomenon (interference of waves) can be attributed to purely conceptual entities like probabilities. The probability of some event involving a particle cannot be interpreted as one of its properties.

The wave function of a free electron is a plane wave of a given frequency; the particle's energy is the Planck constant times that frequency. If it has momentum in a particular direction, the wave front is perpendicular to that direction and the spatial oscillations are perpendicular to it and of a wavelength equal to the Planck constant divided by the momentum. This is not only not localized but is in fact spread over an unlimited distance in the lateral direction. The particle distribution is then spread out in the same way; to use the popular image, there is equal probability of finding the particle anywhere.

We can construct waves of quite complicated form by taking linear combinations of these waves with a common frequency and different momenta. In empty space, waves of given frequency all have momenta of the same magnitude but differing directions. Any linear combination of them describes free electrons, which may then have quite complicated forms. In the case of photons, light signals spreading out from a point source may be characterized in this way; if they have quantized angular momentum, they are characterized as spherical or multipole waves. Thus, waves of given energy are not characterized by a given geometry but only by a given energy or frequency.

It is in this light that we should view the two-slit experiment of Feynman. If the waves are emitted at a given frequency, they will maintain that frequency as they propagate, but the momentum distribution will be determined by the slits and will of course be complicated. However, the wave function will still describe states of a free electron, in the sense that the energy spectrum of possible states will be continuous and arbitrary. It is thus that the electron can be considered as going through both slits. The wave function will not be the sum of two parts, one associated with each slit; it is a single unified description of the quantum state of a single electron. This electron can be detected only in its entirety. The situation cannot be described as one in which the electron goes through *either* one slit or the other. We cannot say then that "there is probability p_1 that it goes through slit 1 and probability p_2 that it goes through slit 2, the sum of p_1 and p_2 being unity." However the probability interpretation may be regarded, it cannot be used in this way; that simply does not correspond to reality. The important thing is that what characterizes the electron is *the wave function as a whole*, which does not have such a form. It is in this sense—that is, because there must be a correlation of the electron densities at different points—that the probability interpretation does not fit normal patterns, where individual events are normally determined in a random way. It is here that the *nonlocality* of quantum mechanics becomes apparent. The states are determined by the patterns of the wave function as a whole and not by what the situation is at any particular point.

This feature is well illustrated by considering the absorption of the particle at the point of the detecting screen where it is detected. The electron will have an initial wave function (that in the presence of the two slits) and a final state depending on its interaction with the detecting screen. Quantum mechanics in principle describes this in terms of a time-dependent combination of initial and final state, in which the initial state fades out with time until it disappears and the final state fades in until that state is physically established. While intuition might suggest that the parts of the initial wave function nearest the point of absorption would disappear first, that is not the way it goes; rather, the *whole initial wave function disappears at the same rate! What is important is the pattern of the wave function as a whole.* The question then remains, what does this imply for the particle, which is supposed to be the underlying entity, and whose probability density only is supposed to be determined by the wave function?

While the symmetry aspects of the problem do not appear in this example, there is another familiar one in which it does—the decay of an atom or a nucleus from an excited state to a lower (ground) state. Consider for simplicity the decay of a hydrogen atom from a p state to the lowest s state. This is a decay from an antisymmetric state to a symmetric one, which requires an antisymmetric perturbation. Since this is decay with emission of a photon, the photon will have to have an antisymmetrical character. This is not a problem for the electromagnetic field, since it is a vector field and therefore has a directionality associated with it. Physically, what happens is that the electron loses angular momentum in the transition process, and the emitted photon carries that angular momentum away. But again, in the decay process, the initial (p-state) wave function fades out uniformly; that is, its pattern remains while its intensity diminishes. Once again, it appears that the electron quantum is best described physically as an extended pattern rather than as a point particle.

The basic physical roots of the issue can be seen by looking at the foundations of the quantum mechanics of fields. "Ordinary" quantum mechanics was formulated in terms of space and time coordinates. This is a direct inheritance from classical physics and its *trajectories*. Dirac showed that such a space–time representation of quantum theory was not necessary, though the determination of his classical dynamical operators was still rooted in it. But these operators are more directly derivable from symmetry considerations than from space–time histories. While some symmetries can be interpreted within a space–time framework, others, it seems, cannot (e.g., the "flavors" of modern particle physics, which we shall discuss in Chapter 33 and which relate to nonclassical properties).

The quantization of the electromagnetic field is the prototype of the problem of field quantization. The coordinates of field theory are not designations of location in space and time but are the amplitudes of the state functions describing the quantum states—the patterns of symmetry of the fields. It is these that describe the dynamics of the system. Consider, for example, the patterns of vibration of a violin or piano string. The string can vibrate only with a discrete set of patterns (each the analog of a musical note). An arbitrary configuration of the string consists of a linear superposition of these patterns, which maintain their individual existence on the string. Each such mode of vibration is characterized by a discrete frequency, which characterizes it as a whole. In terms of mathematical symbols, we can say that the general state can be described by the wave function

$$c_1 \phi_1 + c_2 \phi_2 + c_3 \phi_3 + \cdots$$

where the ϕs represent the various frequency states. We see then that we can specify a given state of the string in either of two ways: one by giving the lateral displacement of the string at each point on it, and the other by specifying the amplitude of each independent mode of vibration, that is, of the c's. In the case of noninteracting systems, each mode is in effect a harmonic oscillator; this is one of the simplest of quantum problems. The quantized energy states are $(n + \frac{1}{2})\hbar\omega$, where ω is the circular frequency, that is, the inverse of the periodic time. Thus, these fields come in quantum units of $\hbar\omega$. The $\frac{1}{2}\hbar\omega$, called the zero-point energy, which is present even in a vacuum, is an embarrassment to the advocates of the point-particle picture, since half-quanta cannot represent half a particle.[41]

Quantum field theory is founded on this second option, that is, that of using as defining coordinates the c quantities rather than the displacements themselves. We analyze the field as a superposition of states of the system as a whole, each of which has its characteristic symmetry pattern.

The same concept may then be applied also to electromagnetic waves or De Broglie's particle waves. It is each wave pattern that is quantized, and each quantum is a nonlocalized rather than a localized entity.

There is a profound difference in outlook here. The point-particle view is founded on the notion of precise position, that is, of the precise specification of spatial coordinates. Quantum field theory, on the other hand, takes as primordial the symmetry patterns of the whole field. This need not then apply exclusively to space–time symmetries but to new nonclassical ones as well. Each of these patterns is a basic quantum state with a definite energy because it is characterized by a definite frequency. Thus, the underlying foundation of quantum field theory is, true to its name, *fields*. Particles become quanta of energy not localized in space or time, which can only interact, or be created or annihilated, in their entirety. Such a view is in precise correspondence with the standard mathematical formulation of the theory, needing no supplementary crutches external to the theory itself to sustain it.

The whole development of modern quantum field theory seems to attest to the dominance of symmetry considerations, which characterize the physics of fundamental particles. It is symmetry that is the dominant characteristic of all particles and their relationships. As noted earlier, many of these symmetries are associated with conserved dynamical variables, though, as we shall see in Chapter 32, new characteristics of matter, with no classical analogs, have had to be invoked, whose only familiar characteristics are in fact symmetries. Again, it does not seem plausible that point particles can carry symmetry information.

This contention stands up to the more sophisticated formulations of ordinary quantum mechanics. Dirac shows that we do not need wave functions at all but can just calculate physical processes and properties (in principle at least) from the commutation rules for dynamical operators. But these operators carry within them the seeds of the symmetries involved in the physics (Noether's theorem). If we go to a space–time representation, they reveal the

---

[41] This was once dismissed as an incomprehensible anomaly of the mathematics. Unfortunately, recent developments have shown that it has important physical implications capable of experimental verification. A whole book has been devoted to it. See P. Miloni, *The Quantum Vacuum*, Academic Press, San Diego, CA, 1994.

space–time implications of these symmetries. All of this seems delightfully uncomplicated, at least in principle.

There remains the question of whether the interpretation of quantum transition processes proposed is or is not consistent with the principle of relativity. No complicated argument is needed to resolve this question; it is to be found in the existence of relativistic quantum theory as developed by Dirac. In fact, the whole of quantum field theory incorporates manifest relativistic invariance. Nonrelativistic quantum theory is simply a low-energy limit of the relativistic theory. Thus, if there appears to be conflict with relativity, it can only have its roots in the imprecision of the nonrelativistic theory in particular situations.

25

The Quantum Mechanics of Multiparticle Systems

There is nothing very complicated about the quantum mechanics of single particles. Mathematical methods permit the simple solution of many problems; for others, approximation methods make it possible to calculate to as high a level of accuracy as we wish. Two-particle problems resolve into two independent motions: that of the center of mass and that of their relative motion. At three particles, difficulties already arise, but this was also true in classical mechanics. With many particles, statistical methods become possible. In between is a difficult range where the direct approach becomes increasingly unwieldy with increasing numbers of particles, and this happens far below the range where statistical fluctuations become negligible. The question of chaos in quantum systems is still quite open.

The following seven chapters will deal with the methods, and the problems, involved in solving a range of many-particle problems, which obviously make up most of the world of physics.

If there were no interactions between particles, many-particle systems would pose no difficulties, since each particle could be treated completely independently of the others.

Problems with two particles also create no particular difficulty. The reason for this is that their motions may be separated into two distinct parts: the motion of the center of mass of the particles and their relative motion. The center of mass may be in motion or not; it is always possible to find a frame of reference in which it is at rest. The deuteron is a nuclear particle in which a neutron and a proton are bound together. The motion of the "deuteron as a whole" is completely separate from its internal constitution and behavior.

Another two-particle system is the hydrogen atom, consisting of one proton and one electron. Here again, the center of mass motion may be decoupled from the internal (relative) motion. Because the proton is some 1840 times heavier than the electron, the center of mass will lie very close to the heavier proton. The relative-motion problem is again effectively a one-particle one. This is just the conventional hydrogen atom problem.

The problem of *positronium* is quite similar to the hydrogen-atom problem, except that the masses of the two particles are identical, in which case the center of mass lies precisely midway between them.

At the three-particle level, three interactions are present (those between each pair of particles). There is no longer a simple way to separate the problem into independent parts. The center-of-mass motion can still be separated out, but two independent but entangled variables remain. An example is the helium-atom problem. Another problem, a bit more complicated, is the hydrogen molecule, in which two electrons cluster around two nuclei.

Even the classical three-body problem is quite difficult technically, and the complete analysis of the solar system quite daunting. The core of the problem is to handle the large number of interactions that correlate the motions of the different planets; the number of these interactions is much greater than the number of the planets themselves. With 10 planets there are 45 interactions! Of course, some are much larger than others, but the theory of chaos makes us sensitive to the fact that the smallest of perturbations may, over a sufficient period of time, so multiply themselves as to make prediction impossible. It is important to make this point to remind us that not all of the problems of quantum mechanics are attributable to that theory alone.

In a quantum-mechanical multibody problem, *particles* are replaced by fields, and we must think of the interaction of many *fields*, which now interact over the whole of their spatial extent. In classical mechanics as in quantum, we are forced to the view that we must treat the system as a whole rather than as a collection of separate parts. If many particles were acted upon by a single force, such a description would be appropriate; when all parts interact directly with each other, we must adopt a more holistic view; independence of particles, or fields, is a myth.

The essential fact is that such complex systems increase rapidly in difficulty with increasing size, and that the complication involved raises problems no less profound than those of trying to comprehend quantum phenomena in terms of classical imagery. We may concentrate on idealized and oversimplified systems because we feel that the familiar is easier to grasp and less paradoxical than the unfamiliar. The truth is, however, that in the one case or the other we ultimately find ourselves faced with unexpected difficulties that cannot be resolved by naive reductionism.

Despite all of this, a great deal of ingenuity has characterized the work of physicists on these problems, and in fact in some respects we have developed more powerful analytic techniques in dealing with multiparticle quantum problems than in, say, the problems of celestial mechanics. In the latter case,

computers have made it possible to program models of hundreds of astronomical objects. But this really corresponds conceptually to the treatment of a
sample of a gas by solving the coupled equations of all its molecules. Here we
find that chaos prevails,[42] making it necessary to forego all information about
individual elements and making new and independent hypotheses of a statistical nature. We have tended to believe that the motions of the objects of the
solar system follow an orderly and predictable path. Almost certainly, however, they too are subject to chaos. If it has not been seen, it is due to the fact
that we have not tracked the system, either experimentally or theoretically, for
a long enough period of time for the chaos to become evident. The time scale
depends on two factors. First is the complexity of the system itself—the more
interacting particles involved, the shorter the time scale. The other, increasing
frequency and/or strength of the interactions, also hastens the onset of chaotic
behavior.

In the realm of intermediate systems—those not complex enough to make
good statistics, but on the other hand sufficiently complex so as to approach
the limits of feasibility of exhaustive calculation—we meet our most difficult
problems. A sort of uncertainty principle operates here between the reliability
of theoretical calculation and the time and effort required to attain it.

As an illustration, let us look again at the planetary system, and simplify it
by ignoring very small bodies and treating only the nine planets. There must
be, inevitably, some uncertainty in specifying their positions and motions at
an initial time. Each will interact with eight others and convey their initial
uncertainties to them; there are 36 such interactions. These in turn react back
again on all. How many chains of three interacting planets are there? Each of
these 36 uncertainties will be communicated to 8 more, to make 288, etc.
Uncertainties are multiplied exponentially. This is for only nine objects, and
neglects the effects on them of their moons and of all the smaller objects in
orbit. Nevertheless, this system is much too small to permit a statistical treatment, for which, as one knows from public opinion polls, even a sample of a
thousand involves an average statistical error of about 4%.

In quantum mechanical systems, the situation is mitigated somewhat by the
fact that a small system has discrete energy states so that small uncertainties
are not permitted. There is a point, however, where the numbers of particles
become so large that the concept of individual particle states becomes meaningless. For instance, in 1 cm³ of a metal, there may be about 10^{23} free elec-

---

[42] The theory of chaos arose from the discovery that a miniscule change in the
initial conditions in the mathematical model of a system could result in radically
altered and unpredictable outcomes. See, e.g., Feynman, *Lectures on Physics*, Sec. 2.6,
Addison-Wesley, Reading, MA, 1965, or Wallace, *Physics: Imagination and Reality*,
pp. 278–280. World Scientific, Singapore, 1991. The time scale on which this happens
is shorter the more complex the system. For the molecules of a gas, for example, it takes
only a minute fraction of a second to lose all initial information.

trons, whose energies are spread over a range of a mere 10 electron volts. That means that the supposed states would be separated by a mere 10^{-22} electron volts. Quantum states, however, cannot be defined to more accuracy than the thermal energy, which is about one thousandth of an electron volt per degree. One can conclude that the individual particle states do not in fact exist, and the system must be treated statistically.

We shall say more of statistical quantum mechanics in Chapter 29.

26

Atoms, Molecules, and the Periodic Table

We consider first the two-electron problem represented by the hydrogen molecule, which illustrates the virtues and hazards of simple pictures. The virtues spring from the fact that we can formulate in simple models ("pictures") the qualitative features of what is going on. The hazards have to do with the fact that the truth may lie not in one or another of alternative views but in a compromise between them.

The hydrogen molecule provides a simple model for the binding of atoms into a molecule. It establishes two principles: (i) The binding is due to the attraction of the electron(s) of one atom to the nucleus of the other, and (ii) in the compound states formed, the paired electrons occupy the same bonding state but have opposite spins, as required by the Pauli exclusion principle.

Quantum theory provides a basis for the explanation of the periodic table of Mendeleev. The quantum states of the hydrogen atom are directly calculable and are found to have recurrences of states of the same type, which roughly correspond to those of the Mendeleev table. In many-electron atoms, problems of correlation through interaction and of particle identity do not permit the accurate assignment of electrons to individual states, but reasonably good approximations can be made by considering each electron to be independently in the field of all the others. Feedback loops are created, however, by the fact that the states for each electron then depend on the aggregate of states of all the others. This is the essence of the Hartree-Fock self-consistent approximation.

Symmetry considerations ensure that the sequence of energy levels obtained, in which energies increase with increasing complexity of both radial and angular states, follows much the same pattern as in the hydrogen atom. But the sequence of states is now occupied sequentially by all the electrons of the atom.

Thus, the periodic table is caused by the recurrence of symmetry patterns in the sequence of energy levels being filled.

Atoms and molecules (and nuclei) have well-defined quantum states, but their calculation is greatly complicated by the mutual interaction of their electrons (or nucleons). One-electron atoms, ions, or molecules pose no particular difficulty. With even two, however, new complications arise. We now have two fields that interact with each other through the electrostatic (Coulomb) force. Chemists characterize them as two charge clouds, a quite appropriate designation. Because they are *identical*, an electromagnetic field (photon) cannot interact with one without interacting with the other. They are also characterized by *spin*; if the atom or molecule is in its lowest energy state ("ground state"), the two clouds will be identical, but the Pauli exclusion principle decrees that their spins will be opposite; thus, their total spin will be zero.

The calculation of their energy states is most easily carried out using the Schrödinger wave equation, though it is in general not solvable analytically; it must be done by approximation or by numerical integration. Good approximations are best found by using physical intuition; not only do they imply a physical picture, but generally the errors implicit in them are evident.

Consider, for example, the hydrogen molecule, which consists of two electrons and two nuclei that are merely protons. There are two obvious ways of looking at this system. According to one, we think of the molecule being formed by bringing together two hydrogen atoms. The atoms interact weakly when they are far apart, but more and more strongly as they approach. The new interactions that come into play as they join are

1.   The mutual repulsion of the two electrons
2.   The attraction of each electron to the opposite nucleus
3.   The mutual repulsion of the two nuclei

There are two attractive forces and two repulsive ones. The trick is to arrange them so as to maximize attractions and minimize repulsions.

The other approach is to consider two electrons in the field of the two nuclei. If we ignore the interaction between the electrons, each may be thought of as being in the lowest state of the electron in the presence of the protons. They would then have to be in the antisymmetric spin state, again because of the Pauli principle.

The first approach is clearly flawed in the following respect: It does not allow the two electrons to be simultaneously close to either of the nuclei. The second (molecular orbital [MO] method) errs in the other direction in taking it to be just as likely for the two electrons to be around one nucleus as for them to be attached to one particular nucleus. Clearly, the mutual repulsion of the two electrons makes this unlikely. So, while each approximation springs from a *picture* of the molecule, they depart from reality in opposite directions. Each illuminates an aspect of the problem: The MO method makes it clear that the lowest energy state is that in which the two electrons have opposing spins, while the other method takes some account of the electronic repulsion. Each approximation can be improved by incorporating features of the other; as each then becomes more accurate, the conceptual difference between them tends to disappear.

But before coming to that problem, we must figure out how to handle the separate elements, the electrons on the one hand and the protons (about 1800 times heavier) on the other. Since the protons have vastly greater inertia, we would expect them to be highly localized while the electrons buzz around them. We may then start by treating the protons as effectively at rest, acting as fixed positive charges. We do not know in advance how far apart their nuclei with their respective charge clouds will be, but we do know that the nuclear positions will remain nearly fixed while the electron states adapt to them. We will, therefore, first solve the problem of the electrons with the protons in arbitrary fixed positions. Once the energy of the ground state of the electrons has been determined as a function of nuclear separation, it will be known how much energy is involved in separating the nuclei by any given amount. This then gives us an effective potential energy of interaction of the protons. Thus, we will have a potential to introduce into the quantum equation of the two nuclei, and their states will be calculable. This is really only a one-particle problem, because the center of mass of the protons can move freely, and only their relative internal quantum state remains to be determined. It is found that at a given separation of the protons, the electron energy will have a minimum; it is then to be expected that the protons will vibrate about this value. But they may also have a rotational energy; and in fact, the vibrational and rotational states interact, because the radial distribution of the proton wave function affects the moment of inertia of the rotational motion.

The separation of the problems of the electronic and nuclear states as described constitutes a quite good approximation that may in principle be extended to all molecules, though of course with ever-increasing difficulty (or rather, labor).

The problem of the electronic states remains. Consider first single atoms. Strictly speaking, the electrons do not have independent states, because of the Coulomb interaction between them as well as the fact that they are indistinguishable. Still, intuition suggests that there is a sense in which each electron has a distinct state in the field of the nucleus and the other electron(s). The trouble with this is that the given electron forms part of the environment of the other electrons to whose presence it reacts and, thus, in a sense, has a part in making its own environment. The question then becomes whether we can find a *self-consistent* solution recognizing this fact. That is to say, Can we find single-electron states so that the field to which each electron responds includes the electron's own effect on all of the others? The answer is that it can be done approximately. The electron may be considered to be moving in the *average* field of the others; what we cannot take account of in this way are specific correlations with other individual electrons due to their mutual interaction. The energy arising in that way, which is known as *correlation energy*, lies outside the scope of an average potential model.

The approximation based on average interaction potentials for individual electrons is known as the *Hartree-Fock approximation*, after its founders.[43] It is widely used in the calculation of atomic states. The calculation involves a process of successive approximation. One makes an educated guess of the average potential, uses this to calculate the atomic states of the electrons, then uses them to *determine* more accurately the field to which the individual electrons are subjected. This process may be repeated to give a still more accurate result.

The principle on which this method is based may be used in other contexts (e.g., nuclear physics) and for more complicated multiparticle problems, like those of collective excitations in solids, which we shall discuss later.

An important point should be made here. Very often, the goal is not to reproduce theoretically with high accuracy atomic energy states verifying those that can be measured directly but rather to provide workable mental images of systems that enable us to draw reliable qualitative conclusions. Recall that Mendeleev discovered regularities in the chemical structure of the elements, which he summarized in a periodic table of the elements. This was based on similarities in atomic states of groups of elements, which led to similarities in chemical behavior, in particular, recurring symmetry properties. These are most easily understood in terms of the concept of electrons occupying single-particle states, the Pauli exclusion principle ensuring that only one particle can occupy a given state. A hierarchy of such states may be calculated, in increasing order of energy, the states being characterized by their energy, total angular momentum and one component thereof, and *spin*, the intrinsic electron angular momentum.

Each of these states has a characteristic symmetry associated with its angular momentum. The states can also be characterized by the number of nodal surfaces[44] of their wave functions; in general, the more nodes, the higher the energy. The labeling of the angular momentum states is quite irrational, based only on a description of observed spectral lines; s = sharp, p = principal, d = diffuse, after which they follow alphabetically: f, g, h, etc. As for radial nodes, the state of lowest energy, labeled 1, has none, that labeled 2 has one,

---

[43] The problem can be formulated as that of finding an operator for producing electronic states in this approximation, which then permits the determination of the states. This is treated in detail in my book *Mathematical Analysis of Physical Problems*, pp. 573–579, rev. ed., Dover Publications, Mineola, NY, 1984. This treatment makes clear the technical aspects of the approximation and shows that the effect of particle identity is incorporated in it.

[44] Nodal surfaces are surfaces on which the wave function is zero. Radial ones are those that surround the nucleus, while the angular ones determine its shape. For instance, p states, with one unit of angular momentum, are zero on a plane through the nucleus. The nodes are the three-dimensional equivalent of the nodes on a (one-dimensional) string, where the vibration with no nodes has the lowest frequency, and those with higher numbers are harmonics.

that labeled 3 has two, etc. The order of the states and the numbers of their radial and angular nodes are given in the following table:

Label	Radial nodes	Angular nodes	Angular momentum units
1s	0	0	0
2s	1	0	0
2p	1	1	1
3s	2	0	0
3p	2	1	1
4s}	3	0	0
3d}	2	2	2
4p	3	1	1
5s}	4	0	0
4d}	3	2	2
5p	4	1	1
6s}	5	0	0
5d}	4	2	2
4f	3	3	3 etc.

Adjacent pairs indicated by curly brackets in the table have very nearly equal energies.

For states with nonzero angular momentum, one of the radial nodes is always at the center (the nucleus).

The factor of spin has to be taken into account. For each *spatial* state, there are two possible spin states for electrons, which we shall call *up* and *down* (with respect to any arbitrarily chosen direction). Since *s* states are isotropic, they are singlets, so *s* states can be filled by two electrons. As for *p* states, there are three independent ones, for their nodal planes may be perpendicular to any one of the three directions in space. These states have *axial* symmetry; they consist of two lobes elongated in one of the three directions. Since they differ only by a rotation in space, they are all of the same energy (i.e., are *degenerate*). Each of these may have two spins, so six electrons may be accommodated in these states. The hydrogen (H) atom has only one electron, which must then be in the 1s state, with its spin, a half-unit of angular momentum $\hbar$ in either direction. The next element, helium (He), has a nucleus of charge 2 and two electrons. These must both be in the 1s state and have opposite spins (and thus, net spin zero).

The next element, lithium (Li), has three electrons, two of which are accommodated in the 1s state and the third in one of the 2s states. Beryllium (Be), with four electrons, will have two 1s electrons and two 2s ones, in each case paired with opposite spins. With boron (B) (5), carbon (C) (6), nitrogen (N) (7), oxygen (O) (8), fluorine (F) (9), and neon (Ne) (10), we fill the remaining 2p states.

Here one begins to see the periodic table emerging, because the electrons of the next eight elements will, in a similar way, fill first the two 3s states and then the following six 3p states. We are now up to 18. The elements of this sequence are similar to those of the preceding $(2s)-(2p)$ sequence and so fall below them in the periodic table:

H	He						
Li	Be	B	C	N	O	F	Ne
Na	Mg	Al	Si	P	S	Cl	Ar

After this, things become more complicated. Potassium (K) and calcium (Ca) use up the two $4s$ states, but after that the $3d$ shell begins to fill. The $3p$ states have, like the $2p$ states, a symmetry axis, but the radial node breaks each lobe into two parts. The angular momentum is about the symmetry axis. In the $3d$ case, there are two units of angular momentum, and a given component can have five different values in terms of $\hbar$: -2, -1, 0, 1, or 2. The wave function has two radial nodes and two angular ones, which makes it a bit complicated to visualize. All of these states have the same energy and symmetry and, again, simply correspond to different orientations in space.

There is, however, no unique way of specifying the five independent spatial states. Since the states all have the same energy, if we make a particular choice (such as the states with the five different values of a component of the angular momentum), independent linear combinations of these are also valid states. We shall conventionally take the states of different quantum number m (z-component of angular momentum) as basis states. These states are physically distinguishable because their different angular momenta give them different magnetic strengths ("magnetic moments") in the z-direction, so that they react differently to a magnetic field in that direction. This gives rise to different energy shifts in the field, so that the degeneracy is destroyed.

The sequence of states $4s$, $3d$, $4p$ contains $2 + 10 + 6 = 18$ electron states in all. The $3d$ states break the pattern of the periodic table and end in krypton, which is, like neon and argon, very stable and inert. A similar following series of $5s$, $4d$, and $4p$ states follow, each member being similar to the corresponding state in the previous series. Thus, germanium (atomic number 32) is analogous to silicon (atomic number 14). However, the energy differences between groups decrease as the principal quantum number (the radial one) increases. Thus, the characteristic properties of members of a family become less sharp; the elements become less stable and chemically active. We shall explore this situation further in chapter 27.

Our primary interest is not in technical details but in qualitative features and concepts, so we shall not press further with consideration of the periodic table. The important point to make here is that, despite resorting to approximations of dubious validity, and which certainly misrepresent qualitative features of the problem, quantum mechanics thus simplified gives us insight into relationships between the properties of different elements. We are not dealing with critical philosophical questions here but with physical features of the real world. It does not add to our insight to dwell on theories of measurement or epistemological niceties or to invoke interpretation or probability or paradoxes. What we are doing is in the classical tradition of physical theories; that is, we are using theory to explore the regularities and common principles governing the features of the real world.

This point will be driven home more cogently still when we turn to the consideration of molecules, where rather literal pictures of molecular structure based on the properties of quantum wave functions will be related to the architecture of molecules and solids.

27

CHEMISTS, PICTURES, AND MOLECULAR ARCHITECTURE

Macroscopic matter, whether solid, liquid, or gas, consists of atoms or molecules. The structure of molecules depends critically on the energy states of the atoms of which they are constructed, always governed by the two principles enunciated in the previous chapter. The geometrical forms of the wave functions involved in the binding of atoms enable us to picture the molecules. To illustrate this point, we discuss the ten elements in the first row of the periodic table, from hydrogen to neon, showing how their binding properties (valences) depend on the energies and shapes of their wave functions. The images invoked depend on viewing the wave functions as charge clouds of very specific geometrical forms. From them, we can deduce the shapes of molecules. In certain cases, we see that the bonds permit certain aggregations of atoms to fit together in such a way as to produce large molecules. Carbon atoms are especially adapted for this purpose, so that an almost infinite number of complex molecules can be produced; DNA is a dramatic example. Carbon is the basis for organic chemistry. Molecules may grow in size indefinitely, to produce solid macroscopic crystals.

These pictorial representations seem to represent an almost tangible physical reality and are so intuitive as to lead us to forget that everything from the guiding principles to the finest details are manifestations of a quantum theory that we have been led to believe is strange and weird and represents not a picture of the real physical world but only our knowledge of the antics of an unpredictable point particle.

What is the mechanism by which atoms combine to form molecules, and what determines when they will do so? The answers lie in the principles of quantum mechanics. Let us start by returning to the simplest possible molecule, that of hydrogen. The molecular orbital (MO) viewpoint described in Chapter 26 provides an explanation. Here, two electrons are put into the field of the two hydrogen nuclei, considered, for these purposes, as point charges. There is a lowest (nodeless) state, or, as the chemists misleadingly call it, *orbital* (though it is, of course, a wave function). But we know that two electrons can be put into the same orbital only if their spins are opposite, as required by the Pauli exclusion principle. Thus, if two hydrogen atoms were to come together to form a molecule, their spins in their atomic states would have to be opposite. If their spin states were the same, one would have to go into a higher-energy orbital (according to the MO model). If we think instead of the alternative model of two atoms whose electrons have parallel spins coming together, the antisymmetry of the electron wave function required by the Pauli principle would effectively create a repulsive force between them.

Two things become clear: First, the attractive force of the second proton is the force binding them and is stronger than the mutual repulsion of the two electrons. Second, the formation of a valence bond takes place between two electrons of opposite spins.

Consider next the bringing together of a hydrogen atom and a helium atom, which involves three electrons. The MO viewpoint again involves limiting the lowest orbital to two electrons, the third having to go into a higher-energy state. While this does not in itself necessarily mean that chemical binding cannot take place, it makes it appear unlikely.

Another argument reinforces that conclusion. Since we are now considering only electrons in the fixed electric fields of the nuclei, we can ask what happens when the nuclei are brought together until they merge to form a single nucleus. In the H–He case, we would arrive at a nucleus of charge 3, surrounded by three electrons. Two of these would be in the $1s$ state, while a third could be accommodated only in the $2s$ state. Since, in the combining atoms, all electrons would be in their ground states, the formation of the molecule appears to be disfavoured. The leftover electron is not available to form a valence bond.

If we turn to the case of two helium atoms, we are dealing with four electrons, only two of which will be in the ground state, and the two others in excited states. The spins will be paired off in both states. The ground states of the separated atoms are both very stable, with tightly bound electrons; two electrons in the combined system will be in higher-energy, excited states. Indications are that a molecule will not be formed; detailed calculations bear that out.

Lithium has one extra electron in a $2s$ state. This extra electron enables it to form a valence bond with a similar electron on another atom. Thus, LiH is a stable molecule.

Beryllium poses a new possibility. At first sight, it might be deemed to be much like helium, since its four electrons fill the $1s$ and $2s$ states, and, no electrons are left for bonding. However, the $2p$ states lie only slightly above the $2s$ states, which gives rise to another possibility: If the last two electrons were in $2p$ states and their spins were parallel rather than opposed, they would be free to form bonds with, say, *two* hydrogen atoms.

Several questions come to mind. First, where is the extra energy to come from to compensate that needed to raise the electrons from $2s$ to $2p$ states? A simple answer is that it comes from the energy of binding of the molecule to be formed. But this raises another question: What about the interaction of two H atoms bonded to the Be electrons? Since they will already be involved in pairing with these Be electrons, and since we have seen that bound pairs repel each other, would this not tend to destabilize the resulting molecule? But now nature has another trick lying in wait to help us, and it can work it in two different ways!

Let us recall the structure of p states; they consist of two elongated lobes, the wave function having opposite signs on the two sides. Now in addition to determining which electron states are available for bonding, the attractive force involved in that bonding will actually exert a pull on the Be electrons, which may result in stretching out the wave functions. There is a simple mechanism for this. If we make a combination of $2s$ and $2p$ functions, because of the different signs of the two lobes of the p function while the $2s$ function has the same sign everywhere and is isotropic, a combination of the two will result in the component functions adding to each other on one side of the nucleus and subtracting on the other—accomplishing precisely the stretching out needed. It must be emphasized that this hybrid wave function is, at least approximately, a quantum state of a definite energy under more complicated influences than in the isolated beryllium atom. The spatially biased forces acting on the electron result in a spatially biased wave function. It is not enlightening to say that the electron has a certain probability of being in an s state and another probability of being in a p state, because the essential feature of the new wave function, its lopsidedness, depends on the interference between the two, which in turn depends critically on relative phases.

If, on the other hand, we formed a combination by subtracting the $2p$ function from the $2s$ function, we would have a wave function stretched out in the opposite direction to the first. This being the case, one hydrogen atom could attach itself on one side and one on the other. This would be done not by promoting two s electrons to p states, but by using one s state and one p state in combination. Thus, at a modest cost in energy, we have found a method of attaching two hydrogen atoms to the beryllium while keeping them at a maximum distance from each other, so that their mutual repulsion is as weak as possible.

Thus, a marvelously ingenious way of using quantum mechanics to overcome the obstacles to molecule building has appeared. This trick will later become a major tool, especially in the case of carbon, where it makes possible the vast range of molecular compounds that result in the marvels of organic chemistry. The additional premium offered by this method offers is that it actually *determines what the shape of the molecule will be!* In the present case, the molecule will be linear, thus, H—Be—H.

Does not this sort of triumph of the theory leave the impression that quantum theory provides a description, in the form of a physical model, of real phenomena rather than some sort of intellectual template of the real world? Chemists do not invoke a sort of shadow world of probabilities when they use theory in this way; nor are they disposed to lament that the uncertainty principle hides from them what is needed to provide physical pictures of chemical processes. Perhaps this is why philosophers, who do not use quantum theory as a working tool to deal with real problems, are more prone to see paradox and incompleteness in it than working scientists do.

This said, let us leave the next element, boron, aside for the present. The situation with carbon is so rich in possibilities that it merits a chapter by itself, so we shall pass on to nitrogen and oxygen, obviously of great practical interest, since they form the air we breathe.

With nitrogen, we reach the point where hybridization is no longer a possibility. The number of bonds that can be made depends on the number of unpaired electrons that remain when five states of the $2s-2p$ group are filled. Since only four spatial states are available, one of them must be doubly occupied by electrons with paired spins. The most obvious candidate is the $2s$ state, and we shall proceed on this assumption.[45] What is left, then, are three unpaired electrons in the three p states. Their wave functions are axially symmetric about the three mutually perpendicular directions x, y, and z, respectively. This dictates the shape of the molecule insofar as the nitrogen atom and its next neighbors are concerned. If hydrogen atoms attach to these three prongs, we obtain the ammonia molecule, which may be described as a somewhat squashed tripod. The nitrogen atom cannot exist stably by itself, with three open bonds; it is too chemically active. It normally exists as the molecule N_2, which is the major component of our atmosphere. Linking two tripods is clearly a rather awkward operation. The best one can do is to make one strong bond, in which two legs join end-to-end; and two other weaker ones, involving p functions with axes of symmetry parallel to each other, complement it. One can imagine another alternative: These weaker bonds could be replaced by stronger ones, four in number, with hydrogen atoms, leading to another molecule called hydrazine (N_2H_4).

In oxygen, two spins must be paired off in one of the $2p$ states. Two bonds remain, in perpendicular directions. Linking them to hydrogen atoms produces water, H_2O. Two oxygen atoms can make an O_2 molecule, with one strong bond and one weak one; or, in parallel with the N_2H_4 molecule, the weaker bond between the oxygen atoms can be replaced by two stronger ones involving two hydrogen atoms to make hydrogen peroxide (H_2O_2).

The purpose of this discussion is not to teach some elementary chemistry

---

[45] The other alternative, since it involves only two p states, would appear to lead to a planar structure and seems never to be realized, presumably because it is energetically disfavored.

but to illustrate practical, down-to-earth applications of quantum theory, to extract it from the realm of the exotic and mysterious into that of traditional science. It is also interesting to note that the theoretical arguments used are based strongly on *approximation* and *intuition* and not abstract mathematical symbolism. They permit us to evoke visual models so primitive as to make us forget that their origin is specifically quantum mechanical. It is difficult, in the light of these considerations, to understand the oft-made statement that "no-one *understands* quantum mechanics." Surely, if this is so, what is implied is a peculiar philosopher's definition of *understanding*. By what criterion, then, can we distinguish what one means by understanding quantum theory from understanding other scientific theories?

28

CARBON AND THE STRUCTURE OF ORGANIC MOLECULES

All life is based on carbon and organic chemistry, which alone makes carbon the most fascinating of all elements. The extraordinary symmetry of its wave functions are such that its atoms can be linked together to fill all of two-dimensional or three-dimensional space. Thus, it appears in nature in two crystalline forms—graphite and diamond. The former consists of layers of hexagonal grids, which are only weakly joined together, thus, its slipperiness, which makes it a good lubricant. Diamond fills all of space with a rigid framework, making it the hardest substance known. Quantum mechanical analysis of their energy level structures explains the blackness of graphite and the brilliance of diamond.

Silicon and germanium, senior relatives of carbon in the periodic table, have some similarities, but these are largely blurred by the fact that the higher energy states that they occupy are less sharply differentiated.

The atomic number of carbon is 6. The ground state should then be $1s^2 2s^2 2p^2$, that is, there should be two electrons in $1s$ states, two in $2s$ states, and two in $2p$ states. This suggests the kind of oppositely directed linear hybrid bonds encountered in BeH_2. These involve only one of the carbon p states, say p_x, hybridized with the s state. These will bind with similarly directed p states of the oxygen atoms.

The two $2p$ electrons should be in the same spin state. This is, of course, possible if they are in a different pair of the three degenerate $2p$ states, which can be designated $2p_x$, $2p_y$, and $2p_z$, whose axes of symmetry are in the mutually perpendicular x, y, or z directions.

Once again, however, the $2p$ states lie so close to the $2s$ states that enough energy of bonding in molecules might permit them to be brought into play.

Consider first, for example, carbon dioxide (CO_2), which, as we know, is a very common gas. Using the principles that we have established, we can deduce its detailed structure.

The fact that there are two oxygen atoms to be attracted to the carbon one suggests that the situation may be similar to that of the molecule BeH_2, where the hydrogen atoms are bonded to hybrid states of beryllium. Here, the states of the carbon electrons are hybrids of a p state of carbon, say p_x, and its $2s$ state, with their extensions in opposite directions. Such states will overlap strongly, and thus bond strongly, with the p states of the oxygen atoms, provided that their spins are antiparallel. In any case, the CO_2 molecule is seen to be linear.

There remain valence electrons in the p_y and p_z states of both carbon and oxygen. It is possible to pair them in weaker bonds—weaker because the wave functions do not overlap in the direction of their extended lobes.

We have, therefore, a qualitative model of the molecule of CO_2, which may be put to experimental test.

If we were to use two of the $2p$ electrons to hybridize with the $2s$ electron, we could make three hybrid states with their axes lying in a plane. If the bonds are identical, they will form angles of 120 degrees disposed symmetrically about the nucleus. This has amazing potential, because with six carbon atoms we can form a hexagonal complex, each atom using two of its hybrids to bond with its two neighbors. The third will have its symmetry axis making an angle of 120 degrees with each of the adjacent sides of the hexagon. If these bond with hydrogen atoms, we get the benzene molecule—a hexagon of carbon atoms with a hydrogen atom attached to each corner. But we may also combine as many of these hexagons as we wish, making a whole hierarchy of planar organic compounds. The ultimate limit of this is a layer of graphite (Figure 8), which may have almost unlimited extent.

We may ask, What of the $2p$-state electrons perpendicular to the plane? Just as in the case of carbon dioxide, these may form weak second bonds. Note, however, that if this takes place, all these electrons will form bonded pairs with opposite spins, so that the whole will, from the viewpoint of conventional chemical bonds, not bond to anything exterior. We know that graphite consists of stacks of these planar layers. At first sight, one might wonder what holds them together. What does is in fact a very weak electrical force; each layer weakly polarizes its neighbors. Aside from this, the layers chemically repel each other, but that repulsion weakens as the overlap of their wave functions becomes insignificant, while the force of electric polarization is of longer range. The result is that the layers are weakly bound and widely spaced. For this reason, the layers slide over each other easily, which accounts for the well-known fact that graphite makes a good lubricant.

An interesting recent development is the discovery that these layers may be bent around to close on themselves, forming spherical complexes (carbon "balls") known as *fullerenes*, named after the famous architect Buckminster Fuller, whose spherical domes made of hexagons (and some pentagons, which relate to the way the hexagons are held together) they resemble.

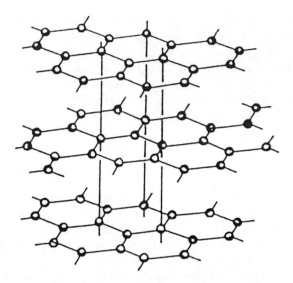

Figure 8.
The crystal structure of graphite, consisting of layers of hexagonally arranged atoms. The bonds between layers are due to weak electrostatic forces, which accounts for the slippery character of graphite.

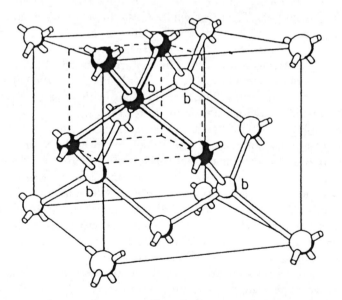

Figure 9.
Basic cell in the crystal structure of diamond (as well as germanium, silicon, and various composite semiconductors). The atom at the center of the basic (dotted) cube will be called a *b* atom; the four at its corners will be designated as *a* atoms. In the composite materials, the *a* and *b* atoms will be of different elements (e.g., InSb, GaAs). The whole crystal can be generated by replicating these cubes indefinitely in the three perpendicular directions in space. (From J.S. Blakemore, *Semiconductor Statistics*, Pergamon, 1962.)

Having seen how carbon can make one-dimensional (linear) structures and two-dimensional (planar) structures capable of an almost infinite number of variations, it remains to be seen how carbon can also make structures to fill space. The elemental structure makes use of all four hybrids possible between one $2s$ electron and three $2p$ electrons. One can envisage it as follows: First imagine a square, and put a carbon atom at its center. Then join the center to two opposite corners at the top and to the two other corresponding corners at the bottom. These are the directions of the four precisely equivalent bonds that one can form; they all make equal angles with each other. Attach a hydrogen atom at the four ends, and you have the molecule of methane, but more miraculously, cubes of this sort may be fitted together to make a crystalline structure of arbitrary extent. By itself, it is a crystal of diamond (Figure 9). The very versatile and technically well-exploited crystals of the elements silicon and germanium have precisely the same structure, as have many compound solids as well.

It must not be forgotten that all of this depends critically on quantum mechanical principles. None of it is conceivable or would be possible if classical Newtonian physics applied on the atomic scale. The example of organic chemistry, the chemistry of carbon, is of particular significance, for it is the basis of the chemistry of life itself.

29

QUANTUM BEHAVIOR IN MACROSCOPIC SYSTEMS: BEYOND WAVE FUNCTIONS

While classical mechanics was from the start believed to be rigidly deterministic, physicists also discovered that it was not realistic to try to predict the properties of a gas by using Newton's equations of motion to determine the trajectories of every elementary particle of matter. So it was that statistical methods were introduced with the goal of determining its macroscopic properties of gases— pressure, density, temperature, etc.—and the relationships between them. J.C. Maxwell played a prominent role in this activity. An ironical feature was that statistics depends on probability and is based on assumptions about a priori probabilities. The assumption made was that, within the constraints of the conservation laws, the distributions of position and velocity of individual particles were completely random; it was assumed that nothing was known of the motions of individual particles.

A similar situation exists in the quantum treatment of the properties of macroscopic systems. We have noted that no physical meaning can be given to individual particle wave functions in solids. Again, statistical methods must be invoked. A striking feature of this approach is that the results obtained for macroscopic properties are again independent of individual particle states. Wave functions are introduced only as basis states in a formalism in which results are independent of the basis chosen. The physical content of the calculations is now revealed to exist in the abstract Dirac operators and their commutation rules. Since these commutation rules are the source of the uncertainty principles between complementary variables, it follows that the uncertainty principles are an essential element in explaining why our physical world is as it is.

In the light of these considerations, the philosophical problems of interpreting the wave function of particles are reduced to ephemeral shadows in the landscape of physics.

We have seen that the electrons in systems of many particles (gases, condensed matter) cannot realistically be said to have quantum states; that is possible only for phase-coherent systems like superconductors. We have noted that individual-particle states cannot exist, even approximately, in incoherent condensed matter systems; they would be merged into a smear of overlapping states, each much wider than their spacing. How then should we describe such systems?

Let us take another point of view. Think first of the atoms or molecules of which condensed matter is constituted. By themselves, they will have different atomic states, but no two electrons in a given atom shall be in the same state. We need not concern ourselves much with the more strongly bound inner states, since, as we have already seen, binding takes place with the outer-shell electrons whose energies lie near the Fermi energy. As we bring the atoms together, the electrons in these states will not remain localized on a single atom. Rather, their energies will be modified by interactions with neighboring atoms. At the same time, once again, because of the Pauli principle, no two will be able to occupy the same state. Their energies will then be spread over a range comparable with the chemical binding energies, which will in general be of the order of several electron volts; this is called an *energy band*. Each energy band corresponds to an atomic state in the isolated atoms. Just as in chemical compounds, two electrons of opposite spin will be accommodated in each spatial configuration.

Depending on the chemical structure of the atoms of the material, the energy spectrum will consist of bands, which will in general be separated by energy gaps. The Fermi energy may fall inside the highest energy band, or between two bands, in which case the gap will lie between a full and an empty band.

As zero temperature is approached, all of the electrons will be in the lowest available states. When the material is subjected to heat, the electrons close to the Fermi energy may be excited if the Fermi energy is inside a band. If the highest occupied state is separated from the lower unoccupied one by a gap, raising the temperature slightly will not cause even the highest-energy electrons to be excited, since there will be no accessible states for them to be excited to. When the thermal energy of these electrons becomes comparable to the energy gap, excitation of the electrons to the upper band will begin to take place. At room temperature, the thermal energy is about one fortieth of an electron volt, while band gaps may be as much as several electron volts.

The two situations envisaged (Fermi energy inside a band or between bands) correspond to different kinds of materials whose properties at low temperatures will be quite different. If the Fermi energy lies between bands, raising the temperature will not result in communicating energy to the electrons; we say then that they have zero, or very small, *electronic specific heat*. (On the other hand, heating may cause the atoms as a whole to jiggle a bit, so that their overall specific heat will not be negligible.) When the Fermi energy lies within a band, the more energetic electrons *will* be able to contribute significantly to the specific heat.

Similar considerations apply when a voltage is applied across the material. Since the electron energy may decrease by moving in the direction of the electric field, the electrons may flow in the direction of the electric field, and the material will be an electric conductor. This will not be the case if the Fermi energy lies in a band gap, since the energy available to the electron from the electric field will be less than the energy needed to excite it into the next higher band. Materials in which this situation prevails will be electrical insulators and will have negligible electronic specific heat.

It is interesting to note that, even in the absence of thermal excitation or electric fields, the electrons near the Fermi surface (usually known as *conduction electrons*) will be able to migrate freely through the material; they will not be localized on individual atoms. But because states with opposite momenta will have the same energy, if one is occupied, the other will be, and no net current will flow.

For example, if the material is at some temperature—let us say, ordinary room temperature—it will have its share of thermal energy, which will be about one fortieth of an electron volt. This will span a very large number of states, among which each electron will be free to move. Electrons will not occupy definite states but will freely exchange energy with each other. That is to say, they will behave more or less like molecules in a gas, in constant chaotic motion due to constant collisions.

We will not then be able to speak of these *dynamic* electrons as being in a fixed quantum state. What we will be able to do, on the other hand, is use statistical analysis to determine how likely it is that a given state will be occupied by some electron or other, given the pool of total energy available. If these *states* are designated $|\alpha\rangle$, we can call the probability of their occupation (averaged over an ensemble of identical systems, for instance), p_α. If we consider some dynamical variable K, its quantum average value in that state will be its diagonal matrix element $\langle\alpha|K|\alpha\rangle$. Thus, the total energy, the sum of that from all quantum states, will be $\sum p_\alpha\langle\alpha|K|\alpha\rangle$, which $\sum$ signifies a summation over all the states $|\alpha\rangle$.

We can then define a *statistical operator* $\sum|\alpha\rangle p_\alpha\langle\alpha|$, which we will designate by Ω. If this operator is expressed as a Heisenberg matrix, it is known as the *density matrix*; we prefer to call it the *statistical operator*, since it is a quantum operator that depends on the statistics of the distribution of energy among particle states.

In many-body systems in which single particles do not exist in quantum states, the statistical operator replaces the wave function or state function as the major working tool. Some of its properties that are easily demonstrated follow:

1. $\text{Tr}\,\Omega = 1$.[46]

---

[46] The *traces* of an operator K is defined as $\Sigma\langle\alpha|K|\alpha\rangle$, where $|\alpha\rangle$ is any complete set of states (called a *basis*).

2. The average value of K in an ensemble of statistically identical states is $\mathrm{Tr}\,\Omega\,K$. It has the remarkable property that *it yields the same result no matter what complete set of basis states is used* to express the trace. It depends only on the quantum operator K. Information about quantum states is found only in the statistical operator.
3. A sort of time-dependent wave equation may be written for Ω, so that its mathematical form can be calculated.

Fano has shown in an article in "Reviews of Modern Physics" that all of these properties can be proven if one starts by *defining* Ω as the operator with the property that the ensemble average of any dynamical operator K is $\mathrm{Tr}\,\Omega K$.

Putting aside the formal theory for a moment, it is important to emphasize the significance of what we have done.

1. We have shown that the problems involved in incoherent many-body systems are distinctly different from those of single-particle or few-particle systems. In the former, wave functions, while not essential, are very useful both in solving problems and in giving intuitive insight into problems. In many-particle systems, on the other hand, the statistical operator is the key to predicting experimental results.
2. Even more strikingly, in the latter case it appears that single-particle wave functions are totally irrelevant, since the measurable average values do not depend at all on the basis states chosen; their role is purely formal. Nevertheless, the whole structure depends on specifically quantum principles. The physical content of the theory is now focused clearly and exclusively on the mathematical representation of the operators for the various dynamical variables.
3. Nothing whatsoever in the statistical ensemble theory justifies using the concepts of trajectories or point particles. On the contrary, all of the underlying physics is that of fields. Field-theoretical techniques can be used to complete advantage.
4. These techniques can be further refined. The theory of *Green's functions*— which provide tools for solving a wide range of problems in various fields of quantum physics, and in particular the theory of the response of systems to external stimuli—reaches beyond and amplifies the statistical operator concept. The purpose of Green's functions is to determine the response of physical systems to external stimuli.

It is important to keep these wider manifestations of the power and validity of quantum theory in mind when confronted with the contrived and oversimplified portrayal represented by the search for paradoxes designed to discredit it.

Nor should it be ignored that quantum behavior at all levels is entirely different from classical, and that the differences are profoundly physical in origin and reflect little of the considerations with which interpretations are concerned.

29.1. Elementary Excitations as Quanta

A bit of operational mathematics is outlined here for those curious about how the idea of elementary excitations of systems is expressed formally. We will use Dirac's bra–ket notation. Quantum states will be designated by symbols such as $|\alpha\rangle$, where α stands for the quantum numbers defining the state. We represent the ground (lowest energy) state as $|0\rangle$. The higher states will be designated as $|n\rangle$, where $n = 1, 2, 3, \ldots$ (We will be concerned here with systems with discrete states, which can therefore be enumerated in increasing order of magnitude.)

These states can be determined by the operator equation $H|n\rangle = E_n|n\rangle$, which is a generalization of the Schrödinger wave equation. H is the operator for the energy; the fact that there are discrete states indicates that the equation can be solved only when the energy has one of the values E_n. We need not be concerned here with *how* the equation is solved; this is a technical matter and depends on the detailed nature of the operator.

Suppose now that another operator K can be found such that $[H, K] = \Gamma K$, where $[H, K] = HK - KH$. Γ is simply a number that is not affected by the operator; that is, $\Gamma(K|n\rangle) = K(\Gamma n)$. If we let the operators operate on the state $|n\rangle$, we get

$$HK|n\rangle - KH|n\rangle = \Gamma K|n\rangle$$

Since $H|n\rangle = E_n|n\rangle$, this may be put in the form

$$H(K|n\rangle) = E_n(K|n\rangle) + \Gamma(K|n\rangle)$$

We can then say that $K|n\rangle$ is another eigenstate of the system with energy $E_n + \Gamma$. It follows that Γ is the amount of energy excitation of the system; we call this an *elementary excitation*. If the system has a lowest energy state, higher energy states may be obtained by repeating the operator K. Another way of putting it is to say that Γ is the quantum of energy of the system, the quanta being created by the operator K.[47]

---

[47] How to *find* the operator K for a given system is another question, requiring in general some physical insight.

30

More on Phases: The Effect of Magnetic Fields

The interactions of electric and magnetic things and the discovery that they are describable in terms of fields has been a source of wonder from the early 19th century to the present day. H.C. Oersted discovered that electric currents produce magnetic fields whose lines of force circle around them, and Michael Faraday discovered that changing magnetic fields produce changing currents of electrons, which in turn circulate around their lines of force. Finally, it was found that electrons in the presence of static magnetic fields circle in orbits about magnetic field lines. How does this affect the quantum theory of electrons? These electron orbits being oscillators, their energies must be quantized. This is the simplest of all quantum problems; the electron energies are simple multiples of a definite quantum energy. The classical orbits, or the quantum wave functions, turn out to have dimensions such that the magnetic field flux through a circle of such dimensions is an integral multiple of a simple "magnetic flux quantum" of magnitude $hc/2e$ (h is Planck's constant, c the speed of light, and e the electron charge). The phase of the wave function changes by 2π as one goes around the magnetic flux lines and changes continually with angle during the cycle.

This magnetic field-induced phase change can be verified in Feynman's two-slit experiment if a weak localized magnetic field parallel to the slits is applied behind the screen. As the magnetic field is varied, the interference pattern will be shifted, going through a cycle as the flux changes by one quantum.

If the field is sufficiently confined, the electron phase will change even if it is never actually in the region of the magnetic field. But the change in phase is determined by another gauge field, which is the source of both electric and magnetic fields and through which they are related. We see then that phases may be important in physical phenomena.

Before engaging in further discussion of many-particle states, whether coherent or incoherent, let us pause to consider the effects of magnetic fields on charged particles such as electrons. A classical charged particle moving in a constant field of magnetic flux **B** is acted on by a force eBv/c proportional to its velocity and perpendicular to its direction of motion (e being the electron charge, v its velocity, and c the velocity of light). It will thus move in a circle about the direction of the field, since in a circle the acceleration is toward its center, whereas its velocity is transverse, that is, tangent to its path. Another way of putting it is to say that it circles around Faraday's *lines of magnetic force*.

The fact that the orbit is circular marks it as periodic; it is easily shown that its frequency is $f = eB/2\pi mc$, m being the electron mass. Using Planck's formula expressing energy quanta in terms of frequency ($E = hf$), we get as the energy quantum for the electron in the magnetic field $E = heB/mc$.

Of course, in quantum mechanics the electrons no longer have orbits but rather wave functions. Now it turns out, not surprisingly, that the spatial dimensions of the wave functions in the direction transverse to the field are very much like the dimensions of the orbit in classical physics. For the lowest state, it can be estimated from the uncertainty principle that if l is the lateral extent of the wave function, l^2 is on the scale of $\hbar c/eB$. The magnetic flux through the classical orbit is then $\pi l^2 B = hc/2e$. This is known as the *flux quantum*; we shall henceforth designate it as Φ. Is it not fascinating that a fundamental characteristic of magnetism should depend on such a simple combination of the three basic constants of nature and that one of them should be the quantum constant h?

Before turning to the question of the effect of magnetic fields on the phases of wave functions, let us consider the nature of magnetic fields themselves. There is something very curious about them, which has to do with our conception of the nature of reality. In the early 19th century, the Danish physicist H.C. Oersted found that a wire carrying an electric current created a magnetic field, thus revealing a deep connection between electricity and magnetism. The lines of force of this field, to use Faraday's image, circulated around the wire. This was quite extraordinary behavior; to find a *mechanical* analog for it taxed the imaginations of physicists of the time, since it seemed that the forces acting between two bodies always acted along the line joining them. Also, the idea of a force depending on the states of motion of the interacting bodies had no mechanical precedent. That the moving electric charges associated with the current in the wire could exert on a magnet a force perpendicular to their motion was puzzling. It was later shown by the American physicist Rowland that the uniform motion of an electrically charged body itself created a magnetic field, again perpendicular to the motion of the charge.

There is an issue here with what we call *reality*. Noting that the magnetic field is created by a uniformly moving charge leads us to argue that if we were moving *with* the charge, it would not appear to us to be moving, and therefore we would not observe a magnetic field. But Einstein's theory of relativity says that the laws of physics are the same in all frames of reference moving with constant velocity relative to each other. Thus, we can toss a ball back and forth in a uniformly moving train or boat, exactly as if we were on terra firma.

We must then ask, How *real* is something whose existence depends on our state of motion and thus on the conditions of observation?

A possible answer is that a charge at rest, although it does not produce a magnetic field, still produces an electric one. One cannot transform away both the electric and the magnetic fields. According to the theory of relativity, electricity and magnetism are different aspects of the same thing, which we can call collectively the *electromagnetic field*. The point of the theory of relativity is just this, that phenomena do not depend on our frame of reference, but that the ways of explaining (or interpreting) them differ when seen from different frames of reference.

From this viewpoint, the reality of the magnetic field is conditional on the frame of reference of our observations, but the electromagnetic field as a whole is real in all frames. Note that the reality encompasses not only the "things," the fields, but all the relations defining them.

We shall discuss later the general question of what we mean by *reality* and how we can establish it. What is at stake here is our reality, the reality of the world of our observations and the laws that give a coherent explanation of them. Whether there is a reality transcending that is not within the scope of natural science to determine and must forever remain inaccessible to verification. It can exist only in our mental conceptions whose relation to natural law, if such there be, must remain beyond our understanding.[48]

One way to verify that magnetic fields affect phases is to use the two-slit experiment for electrons, which we analyzed in Chapter 6. In this case, the beam is split and recombined to create an interference pattern. If we put a very fine magnet behind the slits and parallel to them, we can see whether the interference pattern is affected. It is found that the pattern is displaced by the magnetic field; the more so, the greater the magnetic flux. With sufficient pains, it can even be arranged so that the field is confined to a small enough region that the electrons do not come into direct contact with it. Now, the whole idea of lines of force was that bodies would respond to the magnetic field at the point at which they were located; thus, a tiny magnet would be an indicator of the magnetic field at the place where it was located. In the experiment proposed, however, the effect on the electron phases is strong even when there is no field in the region of the beams themselves; thus, the effect seems to involve an action at a distance. However, the whole thrust of modern physics has been to repudiate that concept by introducing agents mediating the interaction between distant objects; thus, photons carry the electromagnetic force, π-mesons the strong nuclear force, etc. The process is one of emission of a quantum of the mediating field by one particle and its subse-

---

[48] We are suggesting a manifestation of a sort of cosmic Gödel's theorem.

quent absorption by the other. That does not appear to be the case in this instance.

In fact, however, such a field *does* exist and is present even where there is no magnetic field. It is a vector field commonly known as the *vector potential*. It is an example of a more general class of fields of great importance in modern particle physics, known as *gauge fields*. The vector potential has been known to physicists since before the beginning of electromagnetic theory. For Michael Faraday, who was ignorant of mathematics, it was a physical concept[49]; for theoreticians, it was a mathematically useful tool for the calculation of the properties of electromagnetic waves. In the mathematical theory, the energy of interaction of charged particle and electromagnetic field involved its product with the current. The interdependence of electric and magnetic fields made it difficult to formulate the quantum theory of electrodynamics in terms of the fields themselves. The electric and magnetic fields could both be expressed in terms of the electrostatic potential and the vector potential (designated *A*). *The theory of electromagnetic waves could therefore be formulated simply in terms of them.*

The central role of the vector potential is highlighted by the fact that whereas the magnetic field does not directly impinge on the electrons, the vector potential does. Thus, action at a distance can be eliminated from the theory.

It is hardly surprising that it is the vector potential and not the magnetic field that affects the phase of the wave function. At any point in space, the rate of change of the phase with distance in any given direction is the value of the component of the vector potential in that direction. Thus, the phase changes from point to point along each branch of the electron beam. If one calculates the phase change around a path encircling the magnetic field, it will be 2π, since it clearly cannot have two values at a single point. But it may also be shown that the change is simply the magnetic flux through the path. These two conditions determine that the flux is simply the flux quantum Φ introduced earlier.

If there is no magnetic flux through a closed path, there will be no net phase change in going around it.

Since, in the two-slit experiment (Figure 10), the parts of the wave going through the upper and lower slits pass on opposite sides of the magnetic field, the phase will be changing in opposite directions on the two sides. While the phase is increasing along one side, it must be decreasing along the other, resulting in quite different values of the two phases when the two parts of the wave—which, of course, form a single-photon quantum—recombine at the point of interaction on the screen.

---

[49] What we now call the *vector potential* was anticipated by Faraday as a physical concept in his *Experimental Researches in Electricity*, and he explored it experimentally. He called it the "electro-tonic state."

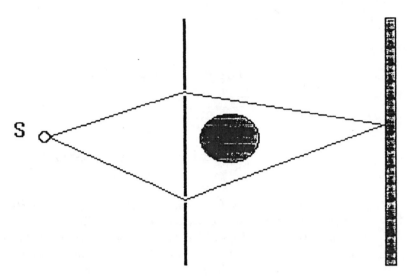

Figure 10.
The two-slit experiment for electrons, with magnetic field. The magnetic field, indicated by the shaded circle, will affect the relative phases of the two branches even though the electrons are nowhere in contact with it.

This may be regarded as an application of Feynman's path integral approach, in which only two paths are possible. Note, however, that the whole interference pattern appears only when we take account of *all* possible paths.

Such behavior was predicted by Bohm and Aharanov in 1956 and was quickly verified experimentally.

When the vector potential was first introduced in the 19th century, it was simply regarded as a mathematical device from which one could determine the real physical entities, the electric and magnetic fields. Quantum mechanics has shown us that it is a physical field in its own right; in fact, it is more fundamental than the fields derived from it. The fact that it manifests itself in a physical experiment whose results have been predicted by the theory gives it a right to the status of a real physical field.

31

Macroscopic Quantum Phenomena

We have called attention to the fact that all of the physical properties of macroscopic matter depend on quantum principles. This is true even of the famous Schrödinger's Cat, which is said to embody a paradox. But we cannot attribute a quantum state to such an object, though we can describe in quantum terms some of the separate chemical reactions going on in it. Both a live cat and a dead cat are in fact a constantly changing assembly of unrelated processes going on in different parts of the organism, as well as taking place between the cat and the world around it. Deadness and liveness are not meaningful quantum numbers.

While this is the normal situation in complex macroscopic systems, the physical world provides examples of macroscopic systems whose components act together as a single quantum state. A number of such states are discussed in this chapter—lasers, superfluids, and superconductors, for example. A more accurate description of the situation is to say that certain materials may be in quantum states of the wholes, where they function as single indivisible macroscopic entities with all of the characteristics that we have defined for microscopic quantum states. A major characteristic is that all components have the same temporal phase coherence in a well-defined quantum energy state.

Type 1 superconductors provide an ingenious device for verifying magnetic flux quantization, based on the fact that magnetic fields cannot penetrate them. A magnetic field is applied to a hollow superconducting tube at a temperature above the superconducting limit. If the temperature is then lowered below the critical temperature, a magnetic field is trapped in the hollow center. It may be verified that the trapped flux is always an integral number of flux quanta.

Brian Josephson showed that if two samples of a superconductor were brought together at a junction, a current could flow without any voltage being applied. The current is due to a phase difference across the junction. This verifies that each superconducting sample is characterized by a single phase.

In another phase-dependent superconducting device known as a SQUID (superconducting quantum interference device), a current in a superconducting wire is split, the two parts separated to form a loop and rejoined further on. A magnetic field is then applied through the loop but not touching the wires, and the current at the far end measured. As the strength of the field is varied, the interference caused by the magnetically induced phase difference between the two sides of the loop causes the strength of the outgoing current to vary continuously.

By a *macroscopic quantum state*, we mean a quantum state of a system at the human scale—that directly accessible to the human senses and formerly thought to be governed only by Newtonian physical laws. From the quantum viewpoint, these states have temporal (phase) coherence; that is, they can be characterized by a single phase. They interact as a single entity. Finally, they manifest behavior unrelated to classical mechanics or electromagnetic theory. In other words, they manifest specifically quantum behavior. Illustrations of this are laser light, superconductivity, and superfluid liquid He_4. We shall briefly consider each of these phenomena, mindful of the fact that they all have their own peculiar subtleties and complications. We shall confine ourselves to the general principles underlying each.

31.1. LASER LIGHT

Most ordinary light is polychromatic; that is, it consists of photons of a spectrum of frequencies. For this reason alone, it has no phase coherence. An important fact that is not universally known or understood is that if a wave has a definite frequency but is of limited length—for example, a finite number of complete waves of a given wavelength—it is not monochromatic. This is because we need other wavelengths (or frequencies) to cause a total destructive interference outside the finite dimensions of the wave. Thus, a periodic wave of finite length cannot consist of a single photon.

If we have light of a definite frequency, it may still be incoherent. For example, the photons may have different directions of propagation, and even when they are parallel they may have different phases. If the beam consists of photons in the same state but with random phases, they cannot reinforce each other because, in random systems, interference effects cancel out statistically. If, however, these photons are all in phase, their fields are additive, so that a wave made of two such identical photons will have twice the amplitude of each. The energy, proportional to the amplitude squared, is then four times that of the original photons. By creating more and more of such identical photons, this multiplication effect can result in monochromatic beams of very high energy. The critical question is, How do we create such high-intensity coherent beams?

The secret lies in a process originally perceived by Einstein (yes, Einstein again!). It is called *stimulated emission*. The way it works is illustrated by imagining that some substance spontaneously emits light at a given frequency due to an electronic transition between two energy levels in the emitting substance whose energy difference is then the energy of the emitted photon. The same material can readily *absorb* light at the same frequency, causing a transition in the opposite direction. Suppose, however, that you have already induced such a transition in the substance, so that it is for a time in this excited state. What then happens if it is further excited by another photon at the same frequency? Einstein used quantum mechanical arguments to show that it would emit another photon with the same phase as the one it absorbed; thus, a pair of coherent photons would be produced. These in turn could stimulate further similar induced emissions, and so on.

It must be noted that this does not imply that quantities of energy can be generated from nothing in this way! Clearly, before one gets laser energy *out* energy has to be invested in stimulating transitions to the excited state; this process is called *optical pumping*. The efficiency of the process can be enhanced by introducing reflecting mirrors at the two ends of the cavity in which the beam is created, to increase the intensity of the stimulating wave. The introduction of the mirrors, on the other hand, has a second effect; that only radiation whose wavelength is such that an integral number of half-wavelengths fit into the length of the cavity can be sustained in it. The frequencies corresponding to these wavelengths are known as *resonant frequencies*. If, however, the cavity is much longer than the wavelengths of the radiation that one is interested in producing, this constraint is not too severe, since there will be a greater number of resonant frequencies per unit frequency range.

It is the phase coherence that makes of the laser beam a single quantum state, with all the consequences that entails.

31.2. Superconductors

Superconductors are not simply very good electrical conductors; the superconducting state of a material is a completely different phase, just as ice, water, and steam are different phases of H_2O. The conductivity is not just enhanced conductivity; it is a state of zero resistance, in which currents can flow almost forever without loss.

The phenomenon was known as far back as 1911, when it was discovered by the Danish physicist Kammerlingh Onnes. It is a phenomenon that, until very recent times, was known to be manifested only at very low temperatures; its discovery then depended on the technology of creating very low temperatures. Onnes had just succeeded in liquefying the gas helium, whose critical temperature was 4.2° Kelvin (i.e., above absolute zero).

A fundamental understanding of the phenomenon was very slow in coming.

The requirements of a fundamental process that could create superconductivity were laid down by Fritz London in the 1930s. Heisenberg at one point offered what he thought was an explanation, but it was quickly recognized that he was wrong. The solution finally emerged in 1957 and was based on an idea of Leon Cooper and formed the basis for a complete theory by Bardeen, Cooper, and Schrieffer. It was John Bardeen's second Nobel Prize in physics; the first had been for transistors.

London's clue was found in a second feature of superconductors, that they were impenetrable to magnetic fields. Try as one might, unless the fields were strong, the lines of magnetic force bent around the superconducting sample and refused to penetrate it. Each superconductor had its critical temperature; this was the temperature below which the material passed from the normal into the superconducting phase. Above the critical temperature, magnetic lines of force inside the material testified to the act of its magnetization. If, still in the presence of the magnetic field, the temperature was lowered below the critical value, the magnetic lines of force were forced out of it. On reheating above that temperature, the magnetic field again penetrated the material. The impenetrability of superconductors to magnetic fields is known as the *Meissner effect*, after its discoverer.[50]

The clue to the Meissner effect is to be found in our old friend the vector potential, the field that circulates around magnetic lines of force and modifies the phase of the wave function from point to point in the presence of, but separate from, the field itself. It was known that the velocity of electrons in a magnetic field, and thus the electric current, contained an extra term, eA/mc. The fact that this circulates around, and is perpendicular to, the field shows that the source of this contribution is the action of the field on the electron's charge. In a superconductor, this motion sets up eddy currents in the surface of the material, whose effect is to generate a field that precisely cancels the applied field, thus shielding the superconductor from magnetic penetration. Fritz London's observation was that the exclusion of the magnetic field penetration could be understood on the assumption that there was an electric current in the material proportional to the vector potential. From this, the exclusion of the field followed by mathematical calculation.

What is the connection between perfect conductivity and the exclusion of the magnetic field? One can rationalize it with a sort of upside-down argu-

---

[50] This may raise questions in the minds of people who have heard of superconducting magnets! The type of superconductor that expels magnetic lines of force is known as *type 1*. There is another sort of superconductor in which, due to impurities, the distance over which there is spatial coherence (coherence length) is less than the penetration depth (the dimension of the thin spatial layer over which the magnetic field can penetrate). For such superconductors (which are called *type 2*), filaments of field may penetrate, embedded in true superconducting material. The superconducting filaments are sorts of vortices, which tend to form, in the superconductor, a regular pattern or lattice.

ment by reductio ad absurdum, which goes like this: Suppose that one had magnetic flux exclusion but anything other than zero resistivity. In that case, the electric currents generated to effect the expulsion of magnetic field would cause a loss of energy, which would be turned into heat. But the superconducting currents have been seen to circulate indefinitely without dissipation of energy. Since the electrons cannot draw energy from the magnetic field, we are led to a contradiction. Thus, magnetic flux expulsion requires zero conductivity.

A classic experiment of W.M. Fairbank and B.S. Deaver in 1961 provided beautiful insights into the the question of phase coherence, flux quantization (as discussed earlier), and the Meissner effect. They constructed a hollow tube of superconducting material, subjected it to a magnetic field when it was above its critical temperature, and then lowered the temperature below the critical value. They found that magnetic flux lines were trapped within the hollow center of the tube; they could not escape because they could not pass through the walls of the tube. The magnitude of the trapped field was then measured and found to be an integral number of flux quanta. The quanta, however, were of amount $hc/2e$, rather than the hc/e originally expected. There is a reason for this, as will become clear when we discuss the theory of Bardeen, Cooper, and Schrieffer.

While the vector potential component of the electron's velocity is identified as the key factor in creating superconductivity, it is also necessary that there be no other component of current to complicate matters. This is ensured by the mechanism proposed by Cooper.

To understand this, it is necessary to recall that in a metal at low temperatures the *free* electrons—those not bound tightly to a given atom—occupy the lowest energy levels available to them. If an electric field is applied, most of these electrons will not be able to respond to it because of the Pauli exclusion principle; we remember that all quantum change is realized by change of quantum state. If the lower levels are occupied, no state will be available for them to move into; they are therefore inert. If, however, an electron is in a state adjacent to the so-called Fermi surface—that is, if its energy is almost at the maximum occupied level—the field may promote it into a level previously unoccupied. Thus, it may move and so carry an electric current. The question is whether there is an *empty* energy level available for it to move into. But the existence of such states is exactly what distinguishes a metal from an insulator and explains why until very recently all known superconductors were metals or alloys.[51]

---

[51] We shall leave aside the ceramic high-temperature superconductors, which have recently attracted attention because of their possible practical importance. The new superconductors are almost certainly the consequence of a mechanism different from the Bardeen-Cooper-Schrieffer one but whose precise nature is not yet clearly understood.

To understand the mechanism, we must clarify some other points. The solid metal may be considered to consist of atomic nuclei surrounded by tightly bound inner-shell electrons, which we will call *ion cores*, and more loosely bound electrons, which in fact can move more or less freely from atom to atom in the crystal; these latter are called *free electrons*. The presence of free electrons is what creates the properties of metals, which, because of them, are good conductors of electricity. However, the ion cores, which form a crystalline array, are not immobile; they interact with their neighbors and are capable of being put into a state of oscillation. Under the influence of their mutual interactions, these motions take the form of wavelike excitations, somewhat like waves on the surface of a lake. These waves, like electromagnetic waves, have quantized energies proportional to their frequencies. These excitations are called *phonons* (suggesting the character of sound waves in an elastic solid).[52]

The last link is to note that the free electrons and the phonons interact with each other. Since the free electrons, being much lighter, move much faster than the ion cores, they may be thought of as leaving a wake of ion core displacement behind them, somewhat like a boat on a lake. Other electrons are affected by this wake, which then provides a mechanism for the interaction of the electrons through the effect of the phonons. (This analogy is not a rigorously exact one but conveys the general sense of the process.)

Cooper and his collaborators were able to show that this process operated particularly strongly between electrons near the Fermi surface (the upper limit of energy of the free electrons) and having opposite momenta and spins. This interaction caused these electron pairs to act in unison with each other and with all other (identical) pairs to form a collective state of depressed energy, the superconducting state.

Since, in the superconducting state, these opposite-momentum pairs (generally known as Cooper pairs) are always occupied simultaneously, and since they are moving in opposite directions, they can carry no net momentum or current. Thus, the only significant electron velocity is that producing magnetic flux expulsion, and London's condition for superconductivity is realized.

A further feature of the superconducting state is that all of the superconducting electrons have the same phase; that is, the superconducting state is a true multiparticle quantum state. This phase coherence is a basic characteristic of superconductors and is the source of an extraordinary phenomenon discovered by a 20-year-old Cambridge undergraduate, Brian Josephson, which won him a share of the Nobel Prize in physics in 1973.

---

[52] When phonons and photons are simultaneously involved in a physical process, it has been jocularly referred to by some physicists as a spectacle of "son et lumière." The first to do so, if I remember well, was Eli Burstein of the University of Pennsylvania.

The Josephson Effect and Schrödinger's Cat. The Josephson effect, which was first predicted theoretically, is this: If two superconducting wires are joined by a thin insulating film (a *Josephson junction*), an electric current can flow between them even though no voltage is applied across it. The strength of this current is proportional to the phase difference between the two samples. But what determines these phases? Apparently, pure chance; thus, it seems that the only way to determine this phase difference is precisely by measuring the Josephson current. This is a quite astonishing and unexpected phenomenon; the more so by virtue of the fact that if a constant voltage is applied across the junction, an alternating current is produced!

The Josephson effect provides striking evidence of the fact that the phases of quantum states play a critical role in macroscopic quantum phenomena, which brings us back again to the question of the adequacy of the probability interpretation of the wave function and verifies the existence of true phase-coherent macroscopic quantum states of matter. In light of this fact, it is worthwhile to consider again Schrödinger's famous cat paradox, in which the cat is said to be in a *mixed state* of half-live and half-dead cat until observed. The superconductor, as well as the other examples in this section, are examples of true quantum states, a requirement of which is phase coherence. We see that this requirement is fulfilled only under quite special and restricted conditions, which apply to superconductors and not to cats.

It is interesting to look at the Schrödinger's Cat problem in the light of this. In a 1988 article in Physics Today entitled "Ask A Silly Question ...", Weisskopf and Feshbach point out that to make a meaningful correspondence between states of cats and states of decaying nuclei one must incorporate the interference of initial and final states. This depends on the relative phases of these states. But it is clear that the states of cats do not provide us with a single phase for either the live or dead state. Rather, each condition involves an enormous (almost infinite) number of separate quantum phenomena, which do not have correlated phases; *a cat, alive or dead, is in this respect not at all like a superconductor!* It is quite unclear what meaning one can attach to such a term as *wave function* for a cat. Quantum states are eigenfunctions of the energy operator. But each part of the cat is constantly exchanging energy with its environment, so enormous numbers of phase-dependent quantum transitions must be taking place constantly. If this distinction is often overlooked, it is because the probability interpretation leads us to forget the essential role of phase.

Superconducting Quantum Interference Experiment. Suppose that a superconducting wire is split to enable it to carry current by two parallel paths. Let us do the same thing with a second sample and then make junctions for each of the two paths, thus creating a superconducting loop. The Josephson current can separate, travel by the two paths, and then recombine into a single wire again. Since the phase differences in the two parallel paths are equal, half the current should flow by each path (Figure 11).

Suppose now that a magnetic flux is passed through the loop between the

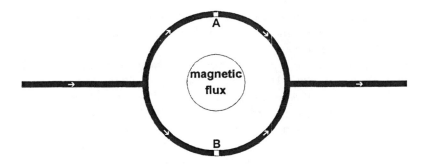

Figure 11.

The SQUID (superconducting quantum interference device). The superconducting Josephson current is split, the two wires on the left side being joined at A and B by Josephson junctions to similar wires on the right. A magnetic flux passes through the loop, creating a magnetically induced difference of phase in the current on the two sides that will depend on the magnitude of the flux.

two paths. This field could be sufficiently localized that it would nowhere come in contact with the wires themselves. Nevertheless, the phases will now differ on the two paths; to find out how much, one has only to integrate the change in phase that is due to the change in the vector potential around the loop, which is simply equal to the magnetic flux through it. The result will be that as the field is varied, the two recombining currents will undergo varying degrees of interference. Each time that the flux through the loop increases by a flux quantum, the current will go through a cycle of change.

An interesting feature of this scheme is that, since the flux quantum is very small (one 10-millionth of a gauss per square centimeter) the interference effect can be used to measure accurately extremely weak fields, or minute variations in fields, of as little as a billionth of a gauss. A device based on this principle is known as a Superconducting Quantum Interference Device (SQUID).

Once again, a specifically quantum effect on the macroscopic scale is seen to owe nothing to the probability interpretation but is a physical manifestation of the phase of quantum states. One must ask, what consequences this has for the probability interpretation and of what use is the *particle* concept in quantum mechanics in throwing light on the phenomenon. How would one undertake to explain it on that basis?

31.3. SUPERFLUID LIQUID HELIUM

Although neutrons, protons, and electrons are Fermi particles—that is, they obey the Pauli exclusion principle—the He_4 atom is a Bose-type particle; any number of such atoms may be put in a given energy state. Both its electrons and its nucleons (neutrons and protons) fill closed energy shells; obviously, any arbitrary number of such atoms may be in, for example, the lowest energy state of the atom. These atoms are completely indistinguishable.

At sufficiently low temperatures, these atoms may condense from a gas into a liquid, called superfluid HeII, occupying a fixed volume but not forming a crystalline array. At absolute zero temperature, they could all be in the lowest energy state; at low temperatures, a large proportion will be in the condensate, but a few will be in low-energy excited states. However, the number of states per unit energy range is small at low energy (it is proportional to the square root of the energy). Below a critical temperature, called the λ point (2.19 K), most of the particles will be in the condensate; above it, almost all will go into excited states.

The nature of the condensation is not like that of water condensing from steam or a liquid condensing into a crystal. These are *spatial* condensations; but in liquid HeII, it is condensation into a coherent momentum state, all atoms having the same phase. Thus, we have again a coherent many-particle quantum state. Among its interesting properties is that it has no viscosity whatsoever, for viscosity arises from internal friction, in which the particles of a liquid exchange momentum, and in the process of which coherent motion is dissipated into heat. In a momentum condensate, this is not possible. The process is somewhat like the stimulated emission that creates coherent laser light; in the superfluid case, the more particles already in the condensate, the more strongly each is drawn into it.

32

COLLECTIVE EXCITATIONS IN SOLIDS

What we call free electrons in a solid are those electrons in the highest occupied energy states, which can migrate freely from atom to atom provided that there are unoccupied states adjacent to the ones they occupy into which they may be excited. These electrons are responsible for most of the electric and magnetic properties of the material. They are subject to various sorts of collective oscillations in which they all move coherently in phase. There is a simple way to determine the frequencies for which this is possible. One first determines how all the components of the system respond individually to an applied field of a given frequency f and wave number vector q, which specifies the wavelength and the direction of propagation of the wave. Since they are charged, their motions create electric and magnetic fields of the same frequency. We then see for what frequencies these fields may be equated to the ones originally applied. This leads to a relation between frequency and wave number, a dispersion relation. Provided this relation can be satisfied with real numerical values, such a wave may be sustained in the material and constitutes a collective macroscopic quantum state. This chapter explores the range of such states.

A crystalline solid is a complex thing and can manifest a wide variety of behaviors, depending on the chemistry of its components, the external forces applied to it, and its interaction with the electromagnetic field, that is, its optical properties. All of these problems can be routinely handled by quantum mechanical techniques, and most properties can be understood only in terms of quantum principles. As we have already remarked, it is inappropriate to attack these problems through the consideration of states of individual electrons, and especially the states of the more energetic (conduction) electrons, which are free to travel through the whole expanse of the macroscopic sample.

Statistical methods must be used, based on the *density matrix* approach outlined in Chapter 29. The problems are not those of charting the course of individual electrons but rather those of determining macroscopic properties such as electrical conductivity, optical refractive index, magnetic susceptibility, and the like. These always involve sums over complete sets of states, the result, however, being independent of the choice of the basis states. Calculations may be much simpler using one set of these basis states rather than another, but these must not be understood to be true quantum states of individual particles.

Our present interest, however, is not in the fact that these macroscopic properties all depend on quantum principles. It is rather to explore the existence of collective (macroscopic, many-particle) coherent quantum states; what we have designated previously as *elementary excitations*. A fascinating *self-consistent* technique permits us to do this. The technique is a broad generalization of a method of solving the problems of complex atoms originally introduced by the British physicist Hartree and the Russian Fock (thus, the *Hartree-Fock method*).

The method proceeds in two steps. The first is to calculate the *response* of the various components of the system (electrons, nuclei, spins, etc.) to an imposed electromagnetic field of definite but arbitrary frequency and wavelength (no relation is assumed between them). This is done using the density matrix techniques referred to earlier. The response takes the form of an electric charge distribution or current, an electric polarization, a magnetization (*magnetic moment*), etc. These responses will, because of the linearity of the problem, also have the same frequency as the applied field. But oscillating electric charges and currents and magnetizations are precisely the sort of things that *create* electromagnetic fields by virtue of Maxwell's electromagnetic equations. Thus, the resulting electromagnetic field may be determined.

In the crucial final step, the field so obtained is taken to be the same as the field originally applied; hence the name *self-consistent*.

The beauty of this technique is that it can incorporate into the quantum state sought a large number of initially separate elements, since the field that acts on all of the components of the system is the net field *created* by all of them.

One can rationalize the method in the following way: It is really a test of whether the system, with its various components, can sustain an electromagnetic field of a certain frequency and wavelength, accompanied by the *coherent* excitation of the various components involved. It will be found under what conditions this *is* possible; in general, this will require a definite relation between wavelength and frequency (a *dispersion relation*).

The whole process can be carried out when the solid is subjected to a constant (fixed) magnetic field; thus, the dynamic properties of the material in a magnetic field may also be determined.

One of the major applications involves the optical properties of the materials, over the whole electromagnetic spectrum. In the presence of magnetic fields, one speaks of the *magneto-optical* properties.

Let us start by describing some of the simplest forms of quantum excitation found in solids.

1. *Phonons* are simply coherent excitations of the ion cores (nuclei plus strongly bound electrons). In *acoustic* (low-frequency) *phonons*, there is only very weak coupling between electrically screened nuclei and *free* electrons, in which case the coupling between screened nuclei and free electrons can be effectively ignored. *Optical* phonons involve strongly ionized ion cores.

2. *Plasma* collective excitations, usually of quite high frequency, are due to the electric (Coulomb) repulsions of the free electrons. Here one first determines how individual electrons react to the arbitrary applied field; this is manifested in oscillating electron currents, which in turn create oscillating electric fields, and these are then identified with the stimulating fields. Thus, one has a gas of electrons oscillating with a specific frequency under the effect of their mutual electric repulsion. Solid-state plasmas, and more complicated magnetoplasmas, attracted much attention in the 1960s and 1970s.

3. *Polarons* are excitations involving the interactions of the oscillation of the ions and the free electrons.

4. *Magnons* are oscillations of spontaneously aligned electron spins in ferromagnetic materials that are due to an external oscillating electromagnetic field. Their reaction takes the form of an oscillating magnetic moment, which then contributes to the stimulating electromagnetic wave, etc. Magnons can thus interact with other excitations to create new *hybrid* coherent states.

5. In the presence of a strong magnetic field, electrons interacting with the electromagnetic field can create fascinating low-frequency electromagnetic waves called *helicons*. These waves, which are characterized by very strong transverse oscillating magnetic fields, can travel with almost arbitrarily low speed; an extraordinary example of a solid material slowing down something that normally has the maximum speed for physical phenomena.

Various hybrid excitations are possible involving the interactions of the preceding forms of quantum excitation. In fact, all components of the solid material may be taken into account. What we have is in effect a method of scanning material to determine what sort of collective excitations are possible at any given frequency.

In cases where self-sustaining monochromatic waves can exist in solid materials, when they are subjected to impinging external waves whose frequency corresponds to the natural frequency, a resonant response can be created in which the wave is strongly absorbed in the solid. In the presence of a strong magnetic field, this gives rise to a *cyclotron resonance*, whose characteristics give considerable insight into the state of the free electrons in the material.

These are only a few examples of a wide range of macroscopically coherent quantum states, in which a macroscopic system can interact integrally with external stimuli. What we are talking about in all cases are quantum states in which the temporal phase coherence of complex systems gives rise to specifically quantum behavior on a macroscopic scale. All of this is far removed from the simplistic paradoxes of contrived systems of single electrons and gives a truer picture of the vast sweep of quantum phenomena.

33

BEYOND THE FRONTIER: NONCLASSICAL PROPERTIES

What is called fundamental particle physics, *or* high-energy physics, *while still being in the realm of quantum theory, takes us into a new realm of reality. It started with the effort to find the ultimate source of all matter, the smallest indivisible elements of which our universe is created. Probing at an ever-smaller scale of size involved using giant machines to produce forms of matter not normally present in our immediate physical world. The search at times seemed to produce so many types of equally fundamental particles that it was difficult to believe that at the most fundamental level one would discover such complexity. But as one sorted the new particles into families, interesting relationships appeared; the use of symmetry properties to classify particles was particularly revealing.*

But as new particles were discovered, so were new properties. That is to say, the characterization of the families involved new mathematical entities for which no intuitively understandable physical image could be found. We had come to a point where interpretation came to a stone wall—no interpretation in terms of familiar concepts was possible. Yet the old phenomenon of complementary and mutually exclusive properties persisted. Thus, particles called K-mesons could not at one and the same time be characterized by parity and strangeness, though what strangeness *meant physically was not resolvable. The vocabulary of the field is now replete with such unimaginable characteristics.*

It seems that if this is where the ultimate realities lie, the 60-year-old argument about the interpretations of wave functions is misplaced. We are left with a mathematical model of a world of which we do not comprehend the basic language.

To this point, we have said little about what is generally called *particle physics*, the search for what are presumed to be the ultimate constituents of our material universe. It is a principle of quantum physics that carrying our investigations to ever-smaller distance scales requires the attainment of higher and higher energies. This is already a curious feature of the quantum world; we are generally inclined to think that when dealing with the very small, everything involved should be appropriately reduced in scale. Yet energy and resources unparalleled in science are required to push our knowledge further and further into the realm of the infinitesimal. It has been pointed out that to reach the energies necessary to test directly our current theories would require accelerators whose dimensions would be greater than those of Earth. Let us recapitulate why this is so. The very small can be investigated only with tools that use very short wavelengths, and quantum mechanics tells us that short wavelengths imply particles accelerated to very high energy. These particles can be produced in sufficient intensity only by an energy input of such magnitude that it must be spread out over large dimensions.

It is natural to ask the question, If we were to find the ultimate particles of which the universe is constructed (if they exist), would all the problems of physics as we know it be solved in principle? The answer is clearly negative. A feature of our standard scheme of things is that the interactions of particles are mediated by other particles. We must remember also that these *particles* are in fact quanta of fields (see Chapter 17). So, clearly, the ultimate picture must incorporate theories dealing with the interactions of these fields. If heavy particles such as the neutron and proton (*baryons*) are made up of quarks,[53] we must study the mechanisms of interaction of the quarks. This leads us to assume the existence of particles (or fields) called *gluons*. Roughly speaking, gluons are to quarks what photons are to electrically charged particles. We are then confronted with the problem of the interaction between gluons. Fortunately, this does not require introduction of a new field; they interact through the same field that ties gluons to quarks. This is well established mathematically but does not invoke any classical images, as does the idea of particles.

But this is not the end of the story. The powerful accelerators that made quarks and gluons accessible also permit the production of a vast array of new product particles, the geneological members of the family of which our familiar world is made. These do not exist explicitly in the world in which we live, being created only ephemerally by the power of our accelerator technology. Current theories (the so-called "standard model") enable us to maneuver with some confidence in this domain. Curiously, however, the internal problems involved in this exercise, to a degree manageable by themselves, exist in a sort of shadow world that is scarcely reflected in the world in which we live and observe. It does not appear to throw any light on natural phenomena

---

[53] See footnote 23 in Chapter 18.

such as, for example, superconductivity. In that sense, the world of our senses cannot be deduced from it.

It is perhaps for this reason that the quantum mechanics of particle physics has led us to invent properties of matter with no counterpart in the familiar world and thus invoke no images in our minds.

A practical question is looming over this matter: What price should our society be prepared to pay to satisfy the curiosity of physical scientists? We do not know how far the process of megaproject construction will have to take us to get the answers we want. We cannot know, until we get there (if we ever do), how far there is to go to reach the desired goal of final answers. The problem becomes one of priorities; of the importance we attach to this particular quest.

Be that as it may, the progress already made has taken us far into unknown country, where the only explanations we can devise are to be found in mathematical formalism. We have made impressive but not conclusive progress on that front but at the cost of foregoing all intuitive meaning to the mathematical symbols and relationships.

Those who read the media accounts—incomprehensible, I am sure, to the layperson, for whom the accounts carry all the conviction of good science fiction—will come across a vocabulary of terms that convey no picture. This vocabulary goes far beyond terms like *quark* and *gluon*, which have at least some analog in our more familiar experience. The problem is not so much with the things we envisage as with the properties we attribute to them. It started with something that Gell-Mann called *strangeness*, named precisely because there was no place waiting for it in our familiar intellectual landscape. There was nothing there to relate it to, except mathematics. Then came *color*, which had nothing whatsoever to do with color as we already know it. Its closest analogy, in fact, is electric charge, though it is much more complicated because it comes in three forms rather than two, so there is no question of a simple $+/-$ relationship as with electric charge. It also carried a curious redundancy: The only real things in physics were in fact colorless! So analogies take you only so far.

If color as a property of equally inseparable quarks and gluons (for once, descriptively named!) is on the edge of familiarity, other properties, analogous to strangeness, are not. They are five in number and complete a complex of six, of which strangeness was the first. Following the pattern that led to the choice of *color* as a descriptive term for the unfamiliar, they are called *flavors*. But what flavors! First we had *up* and *down*, which were constituents of normal (nonstrange) matter. Then came *strange*. Then what appeared to be newer properties appeared. The first was called *charm*, but the theorists could not fit it into a pattern without invoking two other *flavors*, which were initially called *truth* and *beauty*. At this point, it appeared to some physicists that they might be expropriating to themselves too many of the traditional virtues, and *truth* and *beauty* were modestly renamed, without violating alphabetic niceties, *top* and *bottom*. The equivalence of *beauty* and *bottom* did not escape the notice of all. But all this is in fact no joking matter; in a very real sense, physics had entered quite new and foreign territory.

We find ourselves in an Alice in Wonderland world, where we cannot read the signs on the guideposts because they are written in Jabberwocky. All we have to fall back on are analogies that are only formal and mathematical in nature. We scarcely realize how profound a change we are confronted with, so slowly and insidiously has it crept up on us.

Those who are not mathematicians have absolutely no way to understand the *meaning* of these new properties within the general cultural framework of our lives. All we can say is that, to have a consistent mathematical characterization of the physical world, something characteristic of what we consider to be primordial matter must exist, but we cannot understand it at all; we can only give things names. We could just as well have named them A-type properties designated 1, 2, and 3 or B-type properties labeled α, β, δ, μ, σ, τ. The fact is that we have awakened to a new and unfamiliar world where familiar knowledge gives us no clues to behaviors and where the language in which reality is described is unfamiliar to us.

I emphasize this turning point in the history of physics because it seems to me that it complicates the problem of interpretations of quantum mechanics. Here we have stumbled on a territory in which no interpretation, in the sense in which the term is customarily used, is possible. Does this not reinforce the idea that the ultimate interpretation must be found in the structure of the relationships existing in the physical world and not in attempts to fit new experience into the framework of the old, which is what current interpretations attempt to do?

It is not the aim of this book to go into details of the theory of fundamental particle physics, but it is interesting to look at one particular illustration of the point I have just made. It is in fact analogous to other discussions of incompatible sets of variables and is in that sense easy to understand. What is new about it, however, is that it involves a nonclassical variable, *strangeness*, so that there are no simple arguments enabling one to see why the two attributes in question should be complementary.

The problem concerns particles called K-mesons. In their first incarnation, there were two mesonic particles, called τ- and Θ-mesons. They were identical in several respects; they were both boson-type particles, they had the same mass, and both had zero spin and charge. How then could they be distinguished as separate particles? It was a matter of two different modes of decay; the τ-meson decayed into three π-mesons and the Θ-meson into two. The former had a decay lifetime of 4×10^{-8} seconds and the latter 7×10^{-11} seconds. The fact that they decayed in a different way created difficulties because they implied that the two particles were different in respect to another characteristic, called *parity*.

Parity is a matter of symmetry; it has to do with the relation between the particle and its image. By image we mean reflection in a point, which we call the "origin." If the wave function of a particle, which we will call $\Phi(x, y, z)$, is subjected to reflection in a point, it will be turned into $\Phi(-x, -y, -z)$. A particle is said to be of even parity if it is unchanged under the operation of reflection; it is said to be of odd parity if it changes sign under reflection. Now π-mesons were known to have odd parity. But an even number of odd-parity particles would have even parity, and an odd number would have odd parity.[54] If parity is conserved during a transition, this would mean that the Θ-meson had even parity and the τ-meson had odd parity and so apparently could not be the same particle. This puzzle led T.D. Lee and C.N. Yang to ask whether it was true, as had always been believed, that parity was in fact always conserved. They concluded that there was no evidence for this assumption.

When experiments were carried out in which these mesons were created, something quite different was found; two distinct particles still appeared, but this time they varied in strangeness.[55] The values of the strangeness were found to be $+1$ or -1, according to Gell-Mann's theory of strangeness.[56]

So there now appear to be two pairs of neutral particles of the same mass, one of which differed in parity and the other in strangeness. It appeared that they had to be different, but they also had to be closely related! This calls to mind the situation with polarized photons; they could have either definite spins or definite directions of their electric fields, but not the two properties at the same time. In that case, when the photon had a definite value of one variable, it had equal probability of having either possible value of the other. An explanation therefore evolved as follows: The two particles of opposite strangeness, now called K-mesons and designated K^0 and $\underline{K}^0$, have strangeness 1 and -1, respectively. The first is created in a collision of a π^- meson and a proton, producing a K^0 particle of strangeness 1 and a λ-particle of strangeness -1, so that the total strangeness is conserved (at zero). We shall rename the Θ-meson that decays rapidly into two pions K_1 and the τ-meson that decays slowly into three pions K_2. Since the K^0 meson has equal probability of being K_1 and K_2, half of the time it will decay rapidly into a K_1, and the other half of the time slowly into a K_2. Actually, what is created is a beam of K^0 particles so that half of them will decay rapidly (after traveling about 2 cm), while the other half will persist until they have traveled an average

---

[54] This is because the wave function of three independent particles is the product of their separate wave functions.

[55] Since the Θ and τ particles decay only into nonstrange π-mesons, they both have zero strangeness.

[56] It had been postulated that transitions, whether of decay or creation, were strong when strangeness was conserved and weak when it was not. Thus, if two strange particles were produced copiously in a reaction involving nonstrange particles, the new particles would have to have opposite values of the strangeness, which then canceled itself out. This is how particles of strangeness -1 could be identified.

distance of 12 m. Thus, a couple of meters down the beam, all that will be left are K_2s.

Suppose that one now lets this beam collide with protons. Since the K_2 beam has a probability of $\frac{1}{2}$ of being K^0, which has strangeness 1, and $\frac{1}{2}$ $\underline{K}^0$, which has a strangeness of -1, it can, half of the time, then produce a π^+-meson and a λ-particle with strangeness -1. What is astonishing about this, of course, is that as a result of a collision of two nonstrange particles, two λ-particles (strangeness -1 each), a K^0 particle with strangeness 1, and three π-mesons will have been produced. (Of course, this is possible only if the energy of the initially colliding particles is great enough—greater than the sum of the rest energies of all the created particles. But that is no problem for modern accelerators.) That this was all accomplished with strong reactions makes it all the more surprising.

If one finds the preceding rather confusing, one can ignore the details and think of the moral of the story. It is that parity and strangeness are two complementary attributes, so that for K-mesons (kaons) it is impossible to attribute to a particle definite values of parity and strangeness at the same time.

What distinguishes this example of complementarity (and the unfortunately named *uncertainty principle*) is that it involves at least one property (strangeness) of which we have no intuitive feeling whatsoever. Of course, there is always a way of interpreting the situation to make it appear more simple; in this case, we might say that strange particle *is* strange because its parity is undetermined.

The fact that in the situation just considered *strangeness* has crept in when it was not present at the beginning is quite analogous to the situation with polarized light. One can put ordinary light through a filter that blocks out light with clockwise angular momentum, then put the resulting beam through a linear polarizer, thus creating a beam of which half has clockwise angular momentum.

We cannot insist that the natural world conform to the images and concepts with which we are familiar. We have created extensions of our senses that take us into new realms of experience. We cannot without difficulty similarly extend the landscape of our minds to that which defies the patterns of our thought. Still, it can be done. The process is called imagination, which reaches beyond simple rationality into the domain called intuition. One can contend that such a statement consists only of empty words, signifying nothing. What is intuition? *L'intuition est l'intelligence qui fait de l'excès de vitesse.* There are no final answers, no explanations of everything, even in science.

34

Some Reflections on "Reality"

Conventional interpretations of quantum mechanics are based on the premise of Born that the theory is not a description of a physical world—of a physical reality external to us—but rather of what we can learn by observation and experiment (measurement). The reality that exist is a product of observation of our physical environment. But even more strikingly, it affirms that the reality does not exist until we observe it. This sort of subjectivization marks a revolution in thought about science and has philosophical consequences outside the domain of physical science. It is therefore important to ask what it is about quantum theory that calls for such a radical change of view. By looking back at prequantum physics, we can try to discern whether the germs of this philosophical revolution can be detected there. Why was the reality of Faraday and Maxwell's fields not subjected to philosophical questioning? How was action at a distance reconciled with the then-prevalent mechanical view?

Whatever philosophical inferences may be drawn from quantum theory, it is difficult to define a role for science that does not involve belief in the objective reality of the world of our experience, which it is the goal of science to better understand.

What is the underlying process in the progress of physics? Consider first Newtonian mechanics. It involves a model of the physical world, under the constraint of attaining the greatest possible conceptual simplicity in unifying our understanding of the greatest possible range of physical phenomena. Newton's program begins by postulating particles—elementary components of matter—which, under the influence of forces (including those exerted on it by other particles)—respond in well-defined ways, that is, according to specified laws. Does the theory describe "reality"? It has, in itself, little to connect it to our daily experience of the physical world. Can we not say that the conceptual scheme must be *interpreted* in terms of that experience? What is meant here by the word *interpreted?* In the context in which we speak of interpretation of quantum mechanics, it is implied that we must somehow reduce the abstract, idealized conceptual structure to expression in terms of our ingrained structures of thought concerning our direct experience. That is, we want to explain our objective physical theories in terms of our subjective reaction to our experience.

Obviously, the terms *subjective* and *objective* used here need to be made more precise. A theory cast in mathematical form constitutes a basis that can, by the processes of logic, lead to universally accepted conclusions. But these conclusions cannot a priori be assumed to be conclusions about the behavior of the physical world but merely about relations between mathematical entities assumed to be expressive of observable or measurable physical entities. (One makes here the perhaps questionable assumption that logical deduction is a process invariant under the evolution of physical thought; that is, under changes of scientific concepts.) Thus, we designate logic as objective, while our methods or structures of conceptualizing are subjective and constantly changing.

Note, then, that interpretation involves according primacy to the subjective over the objective.

Where does *reality* come into all of this? To answer that question, we must examine the key step in the process, that of identifying the elements of our experience with those of the mathematical structure (theory) that we have created as a basis for describing the physical world. The most natural sense to give to the word *interpretation* is our manner of identifying the abstract mathematical symbols of theory with the concepts that we use to construct descriptions of our experience, in short, of making the link between the objective and the subjective.

Though *reality* cannot be defined either in terms of our experiences themselves or of the mathematical structure of our theories, it is given substance by making workable (verifiable) identifications between the mathematical symbols and the totality of our experience of the physical world. As that experience is broadened and the identifications confirmed, our perception of reality is reinforced.

An obvious question to ask at this point is whether or not this definition of reality is peculiar to physics. Surely, the existence of a mathematical theory is not a condition for our belief in, say, the reality of biological phenomena or of the questions addressed by the (very nonmathematical and nonpredictive) *theory of evolution.* Not even chemistry, which is in a very real sense applied physics, is subjected to the same questioning of *reality* that quantum physics is.

The explanation is surely that our direct experience is, in these other sciences, assumed to be in itself sufficient to define reality. Yet these sciences rest on a foundation of quantum mechanics, a fact that would seem to undermine their status as being based on reality. What distinguishes quantum mechanics as a theory is that it is framed in terms of new and unfamiliar (thus, non-intuitive) concepts and that its manifestations have meaning only within the framework of these concepts.

But is a gene a concept or an element of reality? If the latter, does it derive its reality from its molecular (and thus quantum) origins? Or from the fact that we can play with it, as with a toy? Or is reality defined by the piecing together of the (somewhat disjoint) ways of looking at the world on different scales and in different contexts, so that we must attribute reality to all or to nothing? If this last vision is true, do not our philosophical arguments about quantum theory and reality become academic quibbles about the problems of using our language to describe reality? Can they not then be resolved by adding to our verbal language the language of mathematics?

Insofar as the *experience* of reality is concerned, it must always be incomplete and dependent on our intuition as to what basis for the description of the physical world is most effective, that is, permits the broadest possible verifiable identification of concept with symbol.

But this implies a relative, not an absolute, reality. A belief in the absoluteness of reality depends on the assumption that the incomplete reality, which becomes more and more accessible to us with increasing experience, converges toward an objective reality, which exists independently of our subjective schemes of conceptualization. The motor for this convergence, however, is the constant testing of our theory against observation.

From this point of view, reality is not a function either of our consciousness or of how we intervene in nature (i.e., make a measurement). Physical theory, then, may not be said to represent an absolute reality but rather what, in our present state of understanding, we know about reality. (This begins to sound a bit like Bohr.)

Note that this is not the same as saying that reality exists only in our minds. The whole scientific enterprise, the very existence of science itself, depends on our acting as though an objective reality exists; science then becomes our search for it.

When the process of interpretation sets as its goal the reduction of the objective to the subjective—and this is the story of the interpretations of quantum mechanics—it reverses the normal process of investigation in physical science, which is clearly designed to validate (or invalidate) our intuitive interpretation of experience by testing it against theory. Experiments are normally conceived after theoretical reasoning has either permitted us to predict what their results may be or to determine the value of some theoretically significant quantity. Thus, a measurement of the mass of the Z-particle is thinkable only in the light of a clear understanding of the role of the Z-particle in our theoretical structure.

Let us examine, in the light of the foregoing, the status of classical Newtonian mechanics, since this will enable us to approach quantum mechanics in a more critical spirit. In the simplest of terms, classical mechanics is based on the following assumptions:

- All macroscopic objects in the world are constituted of particles.
- From the laws governing these particles, the laws governing all macroscopic matter can be logically deduced (reductionism).
- The concepts of space and time, which are primordial, are an effective basis for the description of physical processes, as is mass as a property of matter.
- The basic laws (of gravitation and motion) are of the prescribed form. Mass appears in both the laws of motion and the law of gravitation (inertial mass and gravitational mass); these are assumed to be the same.

How complete is the theory; that is, what is the range of phenomena that it can explain? First, it includes everything to do with "ponderable" bodies (i.e., bodies with masses). That it was possible to bring under one theoretical umbrella the fall of the legendary apple from the tree and the orbits of the moon and the planets was impressive evidence of the reality of gravitation.

The reconciliation of the laws of motion with the laws of gravitation required the assumption of action at a distance. Had gravitation not been part of the structure, one might have assumed that it was the action of forces on particles that took place directly, that is, where the particles *were*. So a natural "philosophical" question would be, How do particles interact at a distance? Why should one ask this question? What is the basis for answering it? Surely, it is that all the interactions that could be seen (observed in everyday life) were of that nature. Thus, action at a distance involved a conceptual change. Could it not even be conceived as paradoxical? It might have been assumed (as obvious?) that action at a distance could take place only as a sequence of local interactions (when one rows a boat, the visible action of force is between oar and water, yet the whole boat is propelled). But in the case of gravitation, it could be said that one could not know the intervening agent (just as one cannot know the simultaneous position of position and momentum for a particle in quantum mechanics). Or, alternatively, that there was an invisible agent (ether) through which the force was transmitted. It was not only with the advent of quantum mechanics that strange and marvelous new things, new spooks, crept into physical theory.

At least two major gaps existed in the Newtonian scheme. One concerned the mutual interaction of particles, about which the theory offered no hypotheses to deal with such subjects as the mechanical behavior of macroscopic solid bodies. The other was that, while the theory dealt with ponderable bodies, the world contained things that were not, or seemed not to be, ponderable. The theory had no viable explanation of the blowing of the wind or the propagation of sound, much less of the propagation of light. The urge to describe the physical world in terms of intuitive ideas derived from experience led to the invention of a spooky ether, which was about as elusive as the neutrino at the time of its conception. So far as light was concerned, Newton proposed a particulate theory, but that, too, was elusive and inadequate.

Newtonian mechanics addresses the following problem: Given the state of particles now (i.e., their positions and velocities), where will they be and how will they be moving at any time in the future? To put it differently, what are their trajectories, the smooth, continuous paths that they follow?

In the era of quantum mechanics, it is of course impractical to believe that we can use the linguistic and conceptual framework of intuitions derived from concrete experience to deal with a world in which simultaneous position and momentum cannot even be conceptually defined, much less measured. Subjective paradigm shifts are obviously necessary if we are to enlarge our perception of reality. In specific terms, trajectories must cease to be our preoccupation; the basic program of physics must be broadened.

The Achilles heel of the Newtonian theory was *light*; it was here that things became unstuck. In the meantime, there was an increasing preoccupation among physical scientists with electricity and magnetism. What was taking place was the grafting of these new sciences onto the Newtonian structure as independent phenomena. Gravitation had opened the door to acceptance of action at a distance, though the discovery that magnetism involved forces perpendicular to the line of interaction of bodies was shocking. It was in the end Oersted's discovery of the link between electricity and magnetism that opened the door to radical paradigm change. It led Faraday (intuitively) to a new way of viewing physical phenomena and to new concepts. Maxwell then found the mathematical relationships between these new physical entities (fields), which expressed the correspondence between mathematics and observation. Maxwell's equations constituted the relationship between relationships, which makes a scientific theory of reality. There is an old axiom that the goal of science is to ask the right questions; when this is done, the answers soon become evident. The development of electromagnetic theory provides a good illustration of it.

Theories, however, have creative potential as well, and this gives them their significance. Postulated relationships imply logically derivative relationships. Thus, Maxwell's equations led to the prediction of waves of electricity and magnetism, whose speed of propagation could be predicted by the theory (from parameters, observationally determined, in his equations). It was found that the speed of the electromagnetic waves was essentially the same as the known speed of light, first determined by Römer from astronomical observation late in the 17th century. Thus, a fallout from the theory of electromagnetism was a valid theory of light. It is the sort of situation that leaves no room for doubt that a new reality has been discovered.

It is a reality that goes beyond Newton's program of finding the trajectories of particles. Light does not have position, because waves do not have position. It does have velocity and momentum. Newton himself, not surprisingly, proposed a particulate theory of light; at the time there seemed to be no other possibility. In any case, so far as I know, at no time was it proposed that electromagnetic theory was not causal, as has often been claimed for quantum theory. Thus, causality was not linked to trajectories, and so to the simultaneous prescription of position and momentum. Nor was action at a distance deemed to cause problems with causality. So the scope of physical law, its character, and is goals were broadened without undue problems of interpretation.

Was there a problem of interpretation in the new theory of fields? Faraday, a man without formal mathematical training, had arrived at the concept intuitively, because to him it was the simplest way to visualize what was taking place in the electric and magnetic experiments he was conducting. They were the expression of his intuition. But his inspiration for the idea was visual. For him, electric and magnetic lines of force threaded through space, indicating from point to point the forces operating there. They were the expression of something continuous penetrating the whole of space. The idea was remote from that of particles. In fact, he identified particles with electric charge as the point of convergence, the end point of lines of force; thus, for him, the lines of force, which constituted the field permeating space, were more fundamental than the particle. In this way, he integrated the particle and field pictures, which previously were completely distinct. It must be added, of course, that uncharged particles had the same relation to gravitational lines of force. Faraday's conviction that there was a natural unity in the physical world led him to speculate on a possible connection between electromagnetism and gravitation and thus, by implication, a unified field theory.

Maxwell in his turn supplied the mathematical framework, which permitted the determination of the logical consequences of Faraday's model for the relations between fields. The result was the confirmation of a new reality incorporating the theory and properties of light. It was a gigantic step in our understanding of the physical world.

But the preceding sentence requires further reflection, for it has important implications. What does *understanding* mean if not the revelation of a new level of reality, independent of human thought because it had just been discovered? The fusion of mathematics and concept into objective reality had been realized, however strange and unfamiliar it might be.

Our way of expressing the reality is still intrinsically subjective. Thus, there is a sense in which Bohr is right. Our new theory is, from the conceptual point of view, concerned with the framework we use to communicate our knowledge. But it also describes an external reality that does not depend on the idiom in which we choose to express it, that is, that does not depend on interpretation. This is illustrated in the case of quantum mechanics. The operator formalism of Dirac is a comprehensive mathematical scheme that is a basis for determining all of the measurable properties of physical systems. When the theory is expressed in terms of space and time coordinates—a special and subjective form of the theory—fields emerge. The object of conventional interpretations is to determine what these fields represent physically. Are they real physical things, like electromagnetic or gravitational fields, or simply a code for communicating what we can know about the physical world? The latter implies that there is something in the physical world that we cannot know; this conception surely lies in the domain of metaphysics rather than of physical science. Likewise, the phrase "what we can communicate to each other about physical systems" implies that there exists knowledge about them that we cannot communicate. This is again a purely subjective proposition, so that we are again drawn into the metaphysical orbit and out of the realm of science.

Reality and Other Sciences

Physicists have a tendency to take a reductionist view of the world, in which physics is the base on which all other sciences are founded. While it is a view that may legitimately be questioned, it is interesting to note that its acceptance leads to a curious quandary. Physics has attained a level of theoretical abstraction inconceivable in the other sciences, and this has made possible the philosophical debates that it has engendered. Our probing into the unseen depths of the atomic and subatomic world has removed science further and further from the world of our direct experience and permitted it to rest on layers of inference and extrapolation. Presumably, this has left us with doubts about the meaning of reality. At the level of the more empirical sciences, reality is assumed to be based on our direct experience. That is to say, we define reality in terms of our interactions with the material environment in which we live. We do not question the reality of this immediate world, though it is considered to be in principles dependent on the laws of physics. But if, at its most basic level, we question the reality of the quantum laws on the basis of which we understand the physical world, how can we attribute more certainty to the reality of our direct experience?

May not a solution to this quandary be found in inverting the pyramid and questioning not whether our basic theory in its mathematical form is a valid description of the physical world but whether we have mistakenly read into it a message that has questionable implications outside its intended sphere of application?

Why are these questions about reality, its existence or comprehension, and its nature raised in physics and not in chemistry, biology, geology, etc.? Reduction to physics has taken place in chemistry, to a degree in biochemistry with incursions into molecular biology, but not at all in traditional biology,

Earth sciences, and psychology. But all sciences aside from physics have traditionally been empirical and constructed on a base in which the underlying physics was taken for granted but also deemed of little importance in macroscopic phenomena. Since macroscopic science relies on direct observation and experimentation, there was also no reason for any except the most abstruse philosophers to doubt that they were in direct contact with reality. Things that you could touch, see, hear, taste, or smell were by definition real. While it was generally vaguely thought that reductionism was valid, and that all rested on a physical base, that base was also considered largely irrelevant to the functional pursuit of scientific knowledge.

Reductionism is on much shakier ground nowadays, but this tendency has gone hand in hand with an increasing understanding of some of the internal links within the hierarchy of the sciences (though we tend to agree with Feynman that the concept of hierarchy is inappropriate and that science should be thought of as having horizontal rather than vertical connections). Does this mean that the discovery of means to establish reality in physics carries over into the other sciences that are presumed to rest on it? I think not, and that the argument should be turned around. By any sensible definition of reality, the descriptive sciences are more firmly rooted in it than is physics. All that has happened and is happening in the critical examination of the foundations of physics may be described as extending to physics in a formal way the conviction of reality, which in the more empirical sciences is rooted in our intuitions.

To discover that molecular biology (for instance) is completely reducible to physics would, while strengthening our conviction of the reality of quantum physics, likely raise more philosophical questions about the status of molecular biology. If we start with genes as empirically defined entities whose properties do not require a deeper explanation, we are less likely to evoke these questions. We have centuries of experience with the philosophy of empirical science; it is when science invokes higher levels of theoretical abstraction that new questions are posed.

36

A Final Perspective

In the 20th century, the word science *has had many connotations; it has been portrayed as a doctrine, a religion, a dogma, a resource, and an expertise. Public opinion about it has reflected all of these conceptions, which have aroused a considerable amount of hostility and suspicion. The line between science and technology has been so blurred as to have almost disappeared. All of these developments have sprung from the increasing impact of science on modern society and are reflected in its reification in various forms. But at the basic level of human consciousness, science is none of these things. It is rather a fundamental activity of the human intellect—a process and not a thing. The process is that of finding order in human experience, first, to enable us to cope with the complexities of life, and beyond that, to satisfy the curiosity about all aspects of our conscious existence that is the fundamental characteristic of the human race. Science is the expression of our perception of order in the world, by embodying that order in general principles that enable us to see the relationships of all things. It is that of describing, in the simplest and most general way, how the universe in which we live works.*

Although this definition seems almost trite, public understanding of it would do much to temper public misgivings. A more profound knowledge of our world is not a menace, and the scientific enterprise requires a large measure of humility as constant questioning continues to expose our ignorance.

One of the most striking feaures of our discussion of the meaning of quantum mechanics has been that for over six decades it has been the source of speculation that it represents the emergence of a new kind of science in which human intervention, and the scientific process itself, is believed to have an impact on, if not a key role in, the workings of the physical world. This has in turn aroused skepticism. After all, how could beings who have existed for only the last ten thousandth of the life of our universe and who occupy a small planet circling an undistinguished sun, in a universe of a hundred billion galaxies, each with something of the order of a hundred billion stars, play a decisive role in the mechanism by which it all functions? If this was all made for us, it represents a degree of redundancy beyond the limits of comprehension. At the very least, it is a situation that requires discussion and explanation.

To put the problem in focus, it is useful to start by defining the goals of science and the nature of scientific investigation. A simple proposition from which to start is that science represents our effort to describe the world in which we live. Simple as that proposition is, it has a sobering impact on all of our subsequent discussion. In the first place, it implicitly assumes both that there is a universe external to us that does exist and of which we are conscious. Philosophical comment on quantum mechanics continues to raise the question of reality. There are those in physics who contend that a consequence of quantum mechanics is that it is impossible to know what reality is and that it can be penetrated only bit by bit, as it were, each time we make a measurement. But as Bell has pointed out, measurement-like processes are constantly taking place in the physical world without any intervention by us. In fact, a large proportion of our measurements are made by instruments and have meaning only in the light of pre-existing theoretical models. If these measurements are our only access to reality and derive their significance from the body of previous experience from which we have extracted patterns and relationships (scientific *laws*), in what way does quantum theory differ philosophically from any other verifiable scientific theory? Is there not a conceptual reality defined as the total of what we observe in the physical world and in which we have found or are seeking order and unity? In these terms, reality is the raison d'être of science.

Our efforts to describe the physical universe as something more than a vast array of random data have led us to the formulation of scientific laws designed to encompass a wide range of what were initially independent facts. Knowing Newton's law of gravitation renders it unnecessary to remember the details of the orbits of all the planets. But beyond that, the law, once formulated, leads us to seek constantly other phenomena that can also be explained by the law and, even better, predicted by it.

This process can be observed in the behavior of infants right from birth. The newborn child is an experimentalist seeking to find order in the world to which he is a stranger. One may debate how much memory the infant inherits through his genes, but that is not the central issue. The young child makes instinctive theories built on steadily accumulating experience and experiment in order to make sense of the world. It is a primitive form of the process by which scientists search for simple relationships between complex objects and phenomena. At the end of this rainbow, there is a search for unified theories, to be encompassed in later grand unified theories and ultimately in TOES (theories of everything). It is our recipe for reducing unmanageable complexity to formula.

Lest the preceding discussion be considered too abstract and impractical, it is important to concede that the "theories" of infants are not, like those of Nobel scientists, based on aesthetic considerations and the satisfaction of intellectual conquest. Rather, they have roots in the practicalities of everyday life. There is an appropriate form of theorizing at the diverse levels of scientific experience. Chemists or biologists are unlikely to find much of value in their domain in string theories or even the quarks of the standard model of particle physics. The culture that claims that the process of searching for ultimate components is more fundamental than theories of macroscopic phenomena is based on an unjustified reductionist premise. Watson and Crick's double helix is not more profound by virtue of the latest theories of particle physics. The revolutionary discovery of intrinsic chaos is not illuminated by the theory of gauge fields. But the most striking feature of quantum theory is that its impact stretches over a range of phenomena unprecedented in the history of science.

Can quantum mechanics be characterized as a descriptive theory? Insofar as it is used in normal scientific investigation, that is to say in its uncontroversial mathematical form, it appears to be. In the light of its interpretive and philosophical aspects, it becomes something quite different. One has simply to ask what the debate over interpretation, and the metaphysical speculations generated by it have done to further the process of unifying our knowledge of quantum phenomena. What is the practical impact, for example, of the so-called "many-worlds" interpretation of Everett? How has the point-particle model deepened our understanding of superconductivity? What real phenomenon has been better understood by virtue of the hypothesis of the collapse of the wave function (which does not exist in Feynman's formulation of the theory)? When quantum mechanics is viewed as a description of the physical world, how substantial is the contribution of the 65-year-old debate over interpretations?

Various philosophers have heaped scorn on naive physicists who are satisfied with quantum theory because it works. If it is regarded as a descriptive theory, the most important thing about it is that it works, for that is the function of the theory. When philosophers insist that it is necessary to say why it works, they are asking us to usurp the role of God. If they ask us how it works, the only answer is to be found in the theory itself. Feynman notes that people ask, "How can it be like that?" and he answers, "Nobody knows how it can be like that." The difficulty is that no one can explain what the question means, for it takes us out of the realm of science into metaphysics.

There still remain echoes of the conflict between science and religion, which raged in the late 19th and early 20th centuries. The fallacy behind this dispute was the idea that science and religion were fighting a territorial war for possession of the human soul. The underlying assumption was that science and religion were conflicting established systems of belief requiring either reconciliation by mutual compromise or the extinction of one or the other. Thus, belief in the biblical description of creation clashed with belief in Darwin's theory of evolution. But what drives scientists in the search for simplicity in the natural world is not a fixed system of belief but an open-ended search for more economical ways to organize our knowledge in a framework of underlying principles. The confusion generated in the adversarial atmosphere, however, was understandable to the extent that scientific theories became dogmas and thus obstacles to scientific progress.[57] It is understandable also when science is portrayed as taking on mystical or deep philosophical significance and projects its authority beyond its mandate of description to metaphysical speculation. This remark is not made specifically to disparage such speculation, nor efforts to relate scientific principles to a wider ideological framework, provided that they are identified as speculation and do not claim the authority of science.

Quantum theory has given us a picture of the world full of surprises and insights, one far removed from Newton's clockwork universe. Scientists, and especially physicists, like to attribute aesthetic qualities to their theories. However, the beauty that they find in quantum theory is not based on its imagined mysteries but rather on the fact that it combines underlying simplicity and coherence with revelations of the amazing web of richness and wonder in the universe we inhabit.

---

[57] A by now common saying, originating, I believe, with Peierls in the context of early meson theories, is that evidence in support of the theory was not an important contribution to physics but that evidence contradicting it would be.

INDEX

CPSIA information can be obtained at www.ICGtesting.com
Printed in the USA
LVOW072204050912

297591LV00003B/50/P

9 781461 284680